SWEARING
IS G*OD
F*R YOU

SWEAR!NG IS G✱OD F✱R YOU

THE AMAZ!NG SC!ENCE OF BAD LANGUAGE

EMMA BYRNE

W. W. Norton & Company
Independent Publishers Since 1923
New York · London

For information about special discounts for bulk
purchases, please contact W. W. Norton Special Sales
at specialsales@wwnorton.com or 800-233-4830

Manufacturing by LSC Communications Harrisonburg

ISBN: 978-1-324-00028-0

W. W. Norton & Company, Inc.
500 Fifth Avenue, New York, N. Y. 10110
www.wwnorton.com

W. W. Norton & Company Ltd.
15 Carlisle Street, London W1D 3BS

1 2 3 4 5 6 7 8 9 0

To Team Science Baby, with love and gratitude.

CONTENTS

Introduction:
What the Fuck Is Swearing?

> Swearing draws upon such powerful and incongruous
> resonators as religion, sex, madness, excretion, and
> nationality, encompassing an extraordinary variety
> of attitudes including the violent, the amusing, the
> shocking, the absurd, the casual and the impossible.
>
> Geoffrey Hughes[1]

When I was about nine years old, I was smacked for calling
my little brother a "twat." I had no idea what a twat was—
I thought it was just a silly way of saying "twit"—but that
smack taught me that some words are more powerful than
others and that I had to be careful how I used them.

But, as you've no doubt gathered, that experience didn't
exactly cure me of swearing. In fact, it probably went some
way toward piquing my fascination with profanity. Since
then I've had a certain pride in my knack for colorful and
well-timed swearing: being a woman in a male-dominated
field, I rely on it to camouflage myself as one of the guys.

I

Calling some equipment a fucking piece of shit is often a necessary rite of passage when I join a new team.

So when I discovered that other scientists have been taking swearing seriously for a long time—and that I'm not the only person who finds judicious profanity useful—I was fucking delighted! I first began to realize there was more to swearing than a bit of banter or blasphemy when I happened to read a study that involved sixty-seven brave volunteers, a bucket of ice water, a swear word, and a stopwatch. I was working in a neuroscience lab at the time, and that study changed the course of my research. It set me on a quest to study swearing: why we do it, how we do it, and what it tells us about ourselves.

But what is swearing and why is it special? Is it the way that it sounds? Or the way that it feels when we say it? Does every language have swearing? Why do we try to teach our children not to swear but always end up having to tell them not to swear? Thanks to a whole range of scientists from Victorian surgeons to modern neuroscientists, we know a lot more about swearing than we used to. But, because swearing is still seen as shocking (there was much agonizing about the wisdom or otherwise of using a swear word in the title of this book), that information hasn't made it into the mainstream. It's a fucking shame that the fascinating facts about swearing are still largely locked up in journals and textbooks.

For example, I'm definitely not the only person who uses swearing as a way of fitting in at work. On the contrary, research shows that swearing can help build teams in the workplace. From the factory floor to the operating room, scientists have shown that teams who share a vulgar lexicon

tend to work more effectively together, feel closer, and be more productive than those who don't. These same studies show that managing stress in the same way that we manage pain—with a fucking good swear—is more effective than any number of team building exercises.

Swearing has also helped to develop the field of neuroscience. By providing us with a useful emotional barometer, swearing has been used as a research tool for over 150 years. It has helped us to discover some fascinating things about the structure of the human brain, such as its division into left and right hemispheres, and the role of cerebral structures like the amygdala in the regulation of emotions.

Swearing has taught us a great deal about our minds, too. We know that people who learn a second language often find it less stressful to swear in their adopted tongue, which gives us an idea of the childhood developmental stages at which we learn emotions and taboos. Swearing also makes the heart beat faster and primes us to think aggressive thoughts while, paradoxically, making us less likely to be physically violent.

And swearing is a surprisingly flexible part of our linguistic repertoire. It reinvents itself from generation to generation as taboos shift. Profanity has even become part of the way we express positive feelings—we know that soccer fans use "fuck" just as frequently when they're happy as when they are angry or frustrated.

That last finding is one of my own. With colleagues at City University, London, I've studied thousands of soccer fans and their bad language during big games. It's no great surprise that soccer fans swear, and that they are particularly fond of "fuck" and "shit." But we noticed something

3

interesting about the ratio between these two swear words. The "fuck"–"shit" ratio is a reliable indicator of which team has scored because it turns out that "shit" is almost universally negative while "fuck" can be a sign of something good or bad. Swearing among football fans also isn't anywhere near as aggressive as you might think; fans on Twitter almost never swear about their opponents and reserve their outbursts for players on their own team.[2]

Publishing that research gave me an insight into the sort of public disapproval that swearing still attracts. We were contacted by a journalist from one of the UK's most widely read newspapers. I won't name it, but it's well known for its thunderously moralizing tone while at the same time printing long-lens photographs of women who are then accused of "flaunting" some part of their bodies. We were asked (a) how much money had been spent (wasted) on the research and (b) whether we wouldn't be better doing something useful (like curing cancer). I replied that the entire cost of the research—the £6.99 spent on a bottle of wine while we came up with the hypothesis—had been self-funded, and that my coauthor and I were computer scientists with very limited understanding of oncology, so it was probably best if we stayed away from interfering with anyone suffering from cancer. We didn't hear back. But this exchange brought home the fact that swearing is still a long way from being a respectable topic of research.

Swearing is one of those things that comes so naturally, and seems so frivolous, that you might be surprised by the number of scientists who are studying it. But neuroscientists, psychologists, sociologists, and historians have long taken an interest in bad language, and for good reason.

Although swearing might *seem* frivolous, it teaches us a lot about how our brains, our minds, and even our societies work.

This book won't just look at swearing in isolation. One of the things that makes swearing so fucking amazing is the sheer breadth of connections it has with our lives. Throughout this book I'll cover many different topics, some of which might seem like digressions. There are plenty of pages that contain no profanity whatsoever but, from the indirectness of Japanese speech patterns to the unintended consequences of potty training chimpanzees, everything relates back to the way we use bad language.

Is this book simply an attempt to justify rudeness and aggression? Not at all. I certainly wouldn't want profanities to become commonplace: swearing needs to maintain its emotional impact in order to be effective. We only need to look at the way that swearing has changed over the last hundred years to see that, as some swear words become mild and ineffectual through overuse or shifting cultural values, we reach for other taboos to fill the gap. Where blasphemy was once the true obscenity, the modern unsayables include racist and sexist terms as swear words. Depending on your point of view this is either a lamentable shift toward political correctness or timely recognition that bigotry is ugly and damaging.

What Is Swearing?

Historically, bad language consisted of swearing, oaths, and curses. That's because such utterances were considered to

have a particular type of word magic. The power of an oath, a pledge, or a curse was potentially enough to call down calamities or literally change the world.

These days, we don't really believe that swearing has the power to alter reality. No one expects the curse "go fuck yourself" to result in any greater injury than a bit of hurt pride. Nevertheless, there is still a kind of word magic involved: swearing, cursing, bad language, profanity, obscenity—call it what you will—draws on taboos, and that's where the power lies.

That doesn't mean that swearing is always used as a vehicle for aggression or insult. In fact, study after study has shown that swearing is as likely to be used in frustration with oneself, or in solidarity, or to amuse someone, as it is to be used as "fighting words." That can be a problem: swearing and abuse are both slippery beasts to pin down, and without clear definitions of a phenomenon, how are we supposed to study it? Among the hundreds of studies I've read while writing this book, two common definitions appear over and over again: swear words are (a) words people use when they are highly emotional and (b) words that refer to something taboo. If you think about the words you class as swearing, you'll find that they tick both of these boxes.

More formally, several linguists have tried to pin down exactly what constitutes swearing. Among them is Professor Magnus Ljung of the University of Stockholm, a respected expert on swearing. In 2011 he published *Swearing: A Cross-Cultural Linguistic Study*, in which he defines swearing, based upon his study of thousands of examples and what they had in common, as:

- the use of taboo words like "fuck" and "shit,"
- which aren't used literally,
- which are fairly formulaic,
- and which are emotive: swearing sends a signal about the speaker's state of mind.

In his book *What the F*, Benjamin K. Bergen points out that, of the 7,000 known languages in the world, there is massive variation in the type, the use, and even the number of swear words.[3] Russian, for example, with its elaborate rules of inflection, has an almost infinite number of ways of swearing, most of them related to the moral standing of one's interlocutor's mother. In Japanese, where the excretory taboo is almost nonexistent (hence the friendly poo emoji), there's no equivalent to "shit" or "piss" but, contrary to popular belief, there are several swear words in the language. *Kichigai* loosely translates as "retard" and is usually bleeped in the media, as is *kutabare* ("drop dead"). And, as in so many languages, the queen of all swear words is *manko*, which refers to a body part so taboo that artist Megumi Igarashi was arrested in 2014 for making 3D models of her own *manko* for an installation in Tokyo.

Languages vary in their repertoire of swear words; it's a natural consequence of the differences in our cultures. Bergen suggests that languages fall into one of four classes —what he calls the Holy Fucking Shit Nigger principle. Languages are dominated by either religious swearing, copulatory swearing or excretory swearing. The fourth category refers to slur-based swearing, but so far I haven't come across any languages that are dominated by slurs. There are languages whose most frowned-upon taboos include

animal names. In Germany, for example, you can be fined anywhere from €300 to €600 for calling someone a daft cow, and up to €2,500 for "old pig."[4] Dutch, meanwhile, has a whole host of bad language to do with illness: calling a police officer a cancer sufferer (*Kankerlijer*) can net you two years' incarceration.[5]

Bergen also investigates whether the characteristics of swear words set them apart. In American English, swear words do tend to be a bit shorter than average, but that's not the case in French or Spanish. It's unlikely to be the sound of the words either, as words that sound innocuous in one language can sound grossly offensive in another. This has been played for laughs since Shakespeare's time, with the comedy "English lesson" in *Henry V*. The French Princess Katherine wants to learn English from her maid, Alice. Having mastered "elbow," "neck," and "chin" she asks how to say "*pied*" and "*robe*":

Katherine: "*Ainsi dis-je! 'D'elbow, de nick, et de sin.' Comment appelez-vous le pied et la robe?*"

("That's what I said! 'D'elbow, de nick, et de sin.' How do you say *pied* and *robe*?")

Alice: "'*Le foot,*' Madame, *et 'le count.*'"

Katherine proceeds to have hysterics, the gag being that foot sounds a little like *foutre* and count (Alice's mangled pronunciation of "gown") sounds a bit like *con*:

"'*Le foot' et 'le count.' Ô Seigneur Dieu! Ils sont mots de son mauvais, corruptible, gros, et impudique, et non pour les dames d'honneur d'user! Je ne voudrais prononcer ces mots devant les seigneurs de France pour tout le monde. Foh! 'Le foot' et 'le count'!*"

("'Fuck' and 'cunt.' Oh my Lord! Those are some awful, corrupted, coarse, and rude words, not to be used by a lady

of virtue! I would not say those words before the Lords of France for the world! Foh! 'Fuck' and 'cunt'!")

If we can't judge by the length, the spelling, or the sound of words to tell us what makes a swear word, what can we go on? Some linguists have tried to define swearing by the parts of the brain involved. In his book *Language, the Stuff of Thought*, linguist and psychologist Steven Pinker says that swearing is distinct from "genuine" language and suggests that it is not generated by those parts of the brain responsible for "higher thought"— the cortex, or the brain's outer layers. Instead, swearing comes from the subcortex—the part of the brain responsible for movement, emotions, and bodily functions. It is, he suggests, more like an animal's cry than human language.

In the context of the latest scientific advances, I don't agree. Certainly, swearing is deeply engrained in our behavior, but to read Pinker's definition, you might conclude that swearing is a vestigial, primitive part of our lexicon; something we should try to evolve ourselves away from. There's a vast body of other research that shows how important swearing is to us as individuals, and how it has developed alongside and even shaped our culture and society. Far from being a simple cry, swearing is a complex social signal that is laden with emotional and cultural significance.

!

If we want to define swearing, why isn't it as simple as looking it up in the dictionary? For a start, dictionaries can be incredibly coy about swearing. When he compiled his dictionary in 1538, Sir Thomas Elyot was in no doubt as to the kinds of people who look up dirty words and was having

9

none of it. "If anyone wants obscene words with which to arouse dormant desire while reading, let him consult other dictionaries."[6] Dr. Johnson, on being praised by two society ladies for having left "naughty words" out of his dictionary, replied, "What! My dears! Then you have been looking for them?"[7] At the height of Victorian prudery, the *Oxford English Dictionary* offered "ineffables" for trousers, and well into the twentieth century, while it included all of the religious and racial swear words, it left out "fuck," "cunt," and "the curse." As a side note, I find it interesting that there are plentiful euphemisms for menstruation, including "the curse," "the crimson tide," "Arsenal playing at home," and "having the decorators in," but it has never spawned its own class of curse words. The only ones that I'm aware of are the "bloodclaat" and "rassclaat" in Jamaican patois. In the later part of the twentieth century, other lexicographers were still dropping words based on their acceptability in polite society. In 1976 the American *Webster's* dictionary dropped "dago," "kike," "wop," and "wog," with the foreword note: "This dictionary could easily dispense with those true obscenities, the terms of racial or ethnic opprobrium that are, in any case, encountered with diminishing frequency these days."

The editors of *Webster's* had good motives but were perhaps a little naive. Taking words out of the dictionary doesn't remove them from our language. And while they might have hoped that 1976 marked a new era in racial and ethnic harmony, from the vantage point of forty years on, this seems touchingly optimistic.

So who does get to decide what constitutes a true obscenity? The answer is that we all do. Within our social groups,

our own tribes, we decide what is and is not taboo, and which taboos are suitable for breaking for emotional or rhetorical purposes. Even within the same country, social class can have an effect on what constitutes swearing. According to Robert Graves, author of the 1927 essay *Lars Porsena or the Future of Swearing*, "bastard" was unforgivable among the "governed classes" whereas "bugger" (which Graves can't even bring himself to render in print, preferring to use "one addicted to an unnatural vice" and the oddly xenophobic "Bulgarian heretic") was a much deadlier insult among the ranks to which Graves himself belonged.

"In the governing classes there is a far greater tolerance to bastards, who often have noble or even royal blood in their veins," he wrote. "Bugger" was less offensive among the governed because they "are more free from the homosexual habit," he rather artlessly theorized. But "when some thirty years ago the word was written nakedly up on a club noticeboard as a charge against one of its members," and here Graves can't even bring himself to name Oscar Wilde, "there followed a terrific social explosion, from which the dust has even now not yet settled."

But, while swearing varies from group to group, it still manages to be surprisingly formulaic. So much of swearing, in English at least, uses the same few constructions. For example Geoffrey Hughes, author of *Swearing: A Social History of Foul Language, Oaths and Profanity in English*, points out that the nouns "Christ," "fuck," "pity," and "shit" have nothing in common except that they can be used in the construction for —'s sake.

I thought about the constructions I regularly use and hear and realized that there are many phrases that are

grammatically correct but that are seldom used (and some that are grammatically incorrect, like "cock it" and "oh do cock off" that I use regularly). For example, "shit" is a verb as well as a noun, but I don't think I've ever heard anyone say "Shit it!" or "Shit you," as a complete sentence. "Shit" as a verb currently seems to have a very specific meaning: to wind up or lie to, as in "You're shitting me!" and the charmingly archaically formulaic reply "I shit you not." Meanwhile, the ever-flexible "fucking" and "buggery" can go in almost every swearing phrase.

Common Formulaic Swearing Constructions in British English

	You __	__you	__off	__it	__ing / __y
Cunt (noun)	★	o	o	o	★
Fuck (noun, verb)	★	★	★	★	★
Shit (noun, verb)	★	~	~	~	★
Cock (noun)	★	o	★	★	★
Arse (noun)	★	o	o	★	★
Piss (noun, verb)	~	~	★	~	★
Fart (noun, verb)	~	~		~	★
Bugger (noun, verb)	★	★	★	★	★
Damn (verb)	o	★	~	★	o

Key: ★ "used regularly"
~ "grammatically correct but seldom in use"
o "grammatically incorrect"

The British broadcasting regulator, Ofcom, recently carried out a survey of public attitudes to swearing on TV

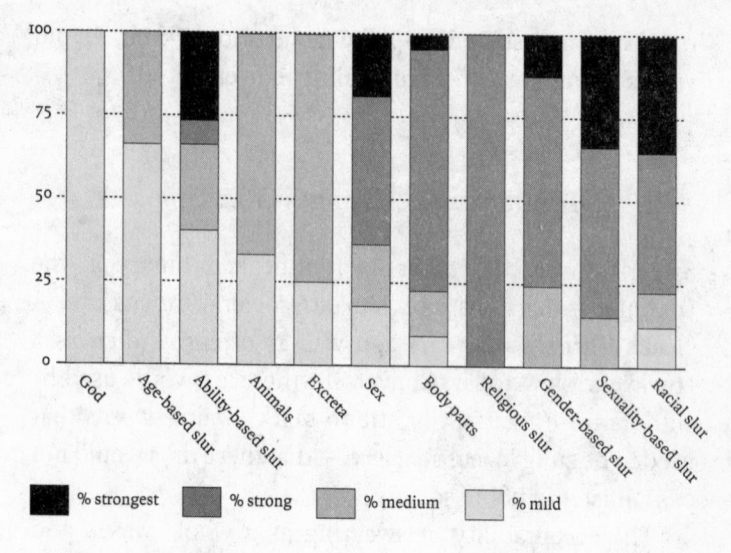

Figure 1: **The proportions of strong and mild swearing by category**

and radio, the results of which I have summarized diagrammatically in Figure 1.[8] Of the "big four" types of swearing in British English (religious, copulatory, excretory, and slur-based), religious swearing was considered the least offensive, while slurs—particularly race- or sexuality-based slurs—were considered most offensive. In fact, a soon-to-be-published study of over 10 million words of recorded speech, collected from 376 volunteers, found that many homophobic and racist slurs have disappeared from people's everyday speech.

Familiar classics like "fuck you" and "bugger off" seem to have been around forever, and they certainly don't lack staying power. Nevertheless, I'm prepared to wager that these swear words will seem as quaint as "blast your eyes"

or archaic as "'sblood" in a few generations' time. As our values change, swearing constantly reinvents itself.

How Swearing Changes Over Time

Swearing is a bellwether—a foul-beaked canary in the coalmine—that tells us what our societal taboos are. A "Jesus Christ!" 150 years ago was as offensive then as a "fuck" or "shit" today. Conversely, there are words used by authors from Agatha Christie to Mark Twain—words that used to be sung in *nursery rhymes*—that these days would not pass muster in polite society.

The acceptability of swearing as a whole waxes and wanes over time. The seriously misnamed Master of the Revels, who presided over London theater in Shakespeare's day, banned all profanity from the stage. That's why the original quarto editions of *Othello* and *Hamlet* contain oaths like " 'sblood" (God's blood) and "zounds" (God's wounds), both of which are cut completely from the later folio edition. By the time a few generations had passed, and "zounds" was a fossil word found only on paper, the pronunciation shifted to "zaunds" and the word lost all connection with its root thanks to the zealous weeding-out of the term from the popular culture of the time.

The censoring of Shakespeare isn't the only evidence we have of changes in what counts as socially unaccep-table language. Linguists and historians have studied trends over the years and identified a huge shift during the Renaissance in Europe. In the Middle Ages, privacy and modesty norms were very different. Talking of bodily parts

and functions wasn't automatically deemed obscene or offensive. But during the Renaissance, those bodily terms began to replace religious oaths and curses as the true obscenities of the time.

That evolution is still unfolding, with terms of abuse that relate to race and sexuality taking on the mantle of the unsayable and disability following behind. That's partly because we're more aware of the effect of a mind-set known as "othering." Othering is a powerful mental shortcut that we've inherited, way back from our earliest primate societies. We all have the subconscious tendency to identify the differences between ourselves and others and to divide the world into "people like us" and "people not like us." We tend to be more generous toward—and more trusting of—the people who are most like ourselves. The problem is that for hundreds of years (at least) the more powerful groups have persecuted and exploited the less powerful. And the words we have for those people in the less powerful groups tend to reinforce those patterns of subjugation, leading to some incredibly powerful emotions. Steven Pinker (as a white male) writing in the *New Republic*, said: "To hear 'nigger' is to try on, however briefly, the thought that there is something contemptible about African Americans."[9]

Your discomfort with the word will depend on your attitude to people based on their race, the same way that your discomfort with blasphemy depends on what you believe about deities. I know I'm the product of my age, class, and upbringing (your average forty-something, middle-class *Guardian* reader), but I definitely find racist epithets and sexuality-based slurs far more uncomfortable than all of the "shits" and "fucks" in the world. I'd much prefer that bodily functions were the source of swearing's power, rather than

somebody's race or sexuality. Without fucking, most of us wouldn't be here, and the scatological unites everybody: in the words of the Japanese author Tarō Gomi, "everyone poops."

Who Swears and Why?

I confess that the swear words I do like, I use an awful lot. I've used them to make people laugh, to cement friendships and to show a side of myself that's "tough" or "ballsy." And, like almost everyone, I've used swearing in pain and frustration, as a way of being funny, or of sending a warning that I'm close to violence. Shortly after I started living in France in my early twenties a man cornered me on my way home one evening and decided to stick his hand up my skirt. Despite not having made any particular attempt to learn French swear words I was astounded with the fluency—and the fury—with which I told him to go fuck himself in the arse, the son of a whore. In just a few weeks of watching films and television in French I'd unconsciously picked up enough foul language to scare away a street harasser.

I'm by no means a special case. While there are some people who insist that they never swear, almost everyone can be pushed into a surprised outburst one way or another (except for a very specific group of stroke patients, whose total inability to swear has helped us to identify the role of emotion in the brain). We do know, however, that men tend to swear slightly more than women, though that gap is narrowing. We also know that left-wingers are more likely to swear on social media than right-wingers,[10] and that swearing really isn't a sign of a stunted vocabulary.[11]

There are two distinct types of swearing that I'll be making reference to throughout this book. Scientists and linguists make the useful distinction between propositional and non-propositional swearing. Propositional swearing is deliberately chosen for effect, and processed mainly in the left hemisphere of the brain for structure, sound, and meaning. Non-propositional swearing is the unplanned, unintended outburst that comes when we're surprised or hurt, and draws more heavily on the emotion-processing parts of the brain. That's not to say that propositional swearing is "left-brained" swearing and non-propositional swearing is "right-brained" swearing: the various parts of the brain have to work together in complex ways that we're only just beginning to understand in order to produce and understand swearing of any kind.

Even those of us who like to avoid propositional swearing are likely to let out a little non-propositional swearing now and again, but lab conditions mean that propositional swearing is more usually studied. Not because it's unethical to shock someone into a bout of swearing (sometimes quite literally); it's just that propositional swearing is much easier to get volunteers to produce on demand.

The Case of the Disappearing Cock and Ass: Notes on Transatlantic Swearing

One of the difficulties I've encountered in writing this book has been the "separated by a common language" factor. So many of the research studies come from North America, New Zealand, and Australia. While some variant of English

is spoken in each of these countries, there's no denying that their swearing habits can be quite different.

The UK, Australia, New Zealand, and the Republic of Ireland have probably the closest affinity. In each of these countries a proud tradition of jocular abuse and a healthy disrespect for authority combine to make for a robust approach to swearing. The United States and Canada, however, are much more uneven in their attitudes to bad language. There are large segments of society that find swearing of any kind deeply offensive, and who are likely to totally reject propositional swearing of all but the mildest kinds.

A Victorian-style sensibility still held sway throughout the English-speaking world well into the twentieth century. Winston Churchill claimed that he was rebuked by one American society hostess for asking for breast meat when offered chicken. According to Sir Winston she replied: "In this country we ask for white meat or dark meat." To make amends, he sent the offended lady an orchid. Being Winston Churchill, he attached a note that read, "I would be obliged if you would pin this on your white meat."[12]

That's not to say that the UK is without its history of prudery, but genetic drift between the two cultures means that dirty words don't always translate directly. In the UK, the request "can I bum a fag" is nothing more outrageous than the request to scrounge a cigarette, but a fanny pack sounds positively gynecological. Animal names, too, show a marked distinction. Our cockerels become "roosters" in Canada and the United States; in the States, our cockroach is usually emasculated to "roach." Here in the UK, however, an ill-treated ass is more likely to end up in a donkey sanctuary than an emergency department.

That said, the plentiful influence of American culture on the rest of the world has made American swearing familiar to most of us. The opposite is less often true. US audiences might appreciate *Downton Abbey* and *Doctor Who*, but neither provides much of a grounding in UK swearing. I've had to explain some of the peculiarities of British English swearing to my North American colleagues on several occasions. Those most often eliciting a degree of bafflement are tosser, wanker, and twat, so for the sake of North Americans reading this, here's my handy guide.

In the UK, pub drinking is highly ritualized. Drinks are bought in "rounds," where one person takes it upon themselves to go to the bar and order drinks for the whole group. Each member of the party is expected to take their turn in this reciprocal drink-buying, and to participate in drinking those rounds. This explains why Brits in large groups tend to get drunk enough to fall over on a regular basis: it's just our way of being polite. Not to participate is—well, I was going to say "not to participate is to mark oneself as an outsider"—in reality, if you want to be thought of as polite not participating in round buying is unthinkable.* So with this in mind, let's meet Adam, Barry, and Chris.

- Adam has forgotten his wallet tonight. He has to borrow some money so he can get a round in. Adam is a tosser.
- Barry has forgotten his wallet but makes no attempt to borrow money. He drinks but doesn't buy a round. Barry is a wanker.
- Chris always "forgets" his wallet, accepts a drink at

---

* At least until the vomiting starts.

every round and then tries to cadge some money for a kebab on the way home. Chris is a twat.

In Defense of Swearing

And therein lies the power of swearing: for all its shock value, swearing is surprisingly subtle. Deployed skillfully, swearing can be cheeky, funny, outrageous, or downright offensive. And when we use swear words, or hear them used, unique things happen in our brains and in our bodies. The use of profanity can help us withstand pain, diffuse stress, bond with our colleagues, and even help us to learn new languages. It's possibly one of the oldest forms of language we have, given how readily other primates have invented swearing of their own, and it turns out that it's fucking useful.

We're often told that swearing isn't big or clever, that it's the sign of a stunted vocabulary or a limited intellect. But I can assure you that swearing can be intelligent and powerful, that swearing is socially and emotionally essential. Not only that, swearing has taught us about our psychology and our societies. And what we've learned—and how we've learned it—is fucking amazing!

I'm evangelical in my defense of swearing, not just on the grounds of freedom of expression, but because swearing is beneficial to us both as individuals and as a species. Because it's so emotive, it's natural to want to tune out swearing; but research proves we should listen more closely when someone swears, because chances are they're telling us something important. So I'm not necessarily encouraging people to swear more, but I do hope you might give it the respect it fucking deserves.

1 ＊ ＊ ＊ ＊ ＊ ＊

The Bad Language Brain: Neuroscience and Swearing

Most of what we know about the human brain comes from trial and error, often more on the error side of that pairing. Some of the biggest breakthroughs in neuroscience have come from investigations no more sophisticated than shoving a finger inside a hole in someone's head, hanging around Victorian insane asylums and, of course, lots of swearing.

By cataloging the functions and the structure of the brain, neuroscience has helped us to understand how and why we swear. It's a two-way street, though: understanding how and why we swear has helped us to reverse-engineer the structure of the brain. Take one of the first and most famous case studies in the history of neuroscience, that of railway foreman Phineas Gage.

One late September afternoon in 1848, Phineas Gage was hard at work blasting rock faces apart, deep in the heart of Vermont. By all accounts he was hardworking and popular, a man who thrived in the American railroad boom

of the 1840s. His bosses thought he was the most efficient and capable man in their employment and they described him in their reports as being "very energetic and persistent." But it was this energy and persistence that was to change the course of Gage's life: in one decisive moment he went from railroad pioneer and model contractor to sideshow attraction and medical marvel.

Gage's team were busy drilling holes in the rock face so they could blast a path for the railway. The process was a dicey one: first the hole was drilled, then it was filled with explosives and a fuse. Finally, sand was poured on top of the explosives so that everything could be "tamped down"—compacted with a meter-long, six-kilogram rod of iron. No one quite knows what went wrong that day, but as Gage drove his tamping iron into the hole it seems to have caused a spark that detonated the blasting powder and shot the metal rod straight through his head and a further twenty-five meters before it finally landed.

The first doctor on the scene, Dr. Edward H. Williams, later wrote that the damage was so bad that he could see the gaping hole in Gage's head even before he stepped out of his carriage, "the pulsations of the brain being very distinct," he wrote. You'd expect someone with a head wound of that magnitude, at best, to be sitting very quietly feeling sorry for himself but Phineas Gage was—according to Dr. Williams and several of Mr. Gage's colleagues—sitting up and chatting with his workmates, regaling them with the details of the accident.

"Mr. Gage persisted in saying that the bar went through his head," wrote Williams. At first, the doctor didn't believe the story, thinking instead that Gage had been hit in the face

with a flying lump of rock. But then, Mr. Gage "got up and vomited; the effort of vomiting pressed out about half a teacupful of the brain which fell upon the floor." That's quite an evocative picture, even if "half a teacupful" isn't the most rigorous unit of measurement. And even after that incident, he still remained very much awake and alive.

Perhaps the most interesting thing about this accident is not that Gage survived, but that this railway foreman became an essential part of the emerging debate on the structure of the human brain. Gage's accident occurred during a monumental shift in how people thought about the brain, when a debate was raging between those scientists who believed that the brain was like a trifle and others who thought it was more like a blancmange. To explain this dessert-based metaphor, the "blancmange" theory (not an official name) held that our brains are an undifferentiated mass. Each bit is just like the other bits—like a blancmange. But the "trifle" school of thought held that the brain is made up of different parts, each one with a different role to play. If you take away a third of a blancmange you still have blancmange. If you take away one of the layers of a trifle you end up with something even more depressing than trifle. Knowing what we do about brain structure these days, it might seem astounding that the question was ever up for debate, but in 1848 there was no means of scanning the brains of living people, and not many survivors of brain injuries that could be observed in close detail, so the debate raged on.

Most of what we know about Gage's condition results from the observations of Dr. John Martyn Harlow, who took over the case. He wrote two papers that describe in detail Gage's injury and the aftermath: the compellingly titled

"Passage of an Iron Rod through the Head" and the equally inventively named sequel, "Recovery from the Passage of an Iron Bar through the Head."

Harlow had an inquisitive mind, a strong stomach, and what must have been a *really* persuasive bedside manner because he left notes explaining how he literally got inside Gage's head to try to figure out—by touch alone—the exact shape of the damage done by the bar. He wanted to check whether there were any bits of bone or shrapnel in the wound and so, around three hours after the accident, he decided to use his finger:

"I passed in [to the hole in the top of Gage's head] the index finger its whole length, without the least resistance, in the direction of the wound in the cheek which received the other [index] finger in like manner."

It's a striking image: Harlow poking one finger up through the hole in Gage's cheek and another through the hole in the top of Gage's head like a pair of Chinese finger cuffs. This was only the first of many painstaking measurements he made. Over the coming years Gage would be sketched, have plaster casts made of his head, and be measured countless times. Harlow eventually concluded that Gage's left frontal lobe was destroyed (that fateful half a teacupful) but that the right side was completely intact.

We now know that Harlow's description of Gage's injury was spot on. His observations were confirmed in 2004 when doctors from the Brigham and Women's Hospital in Boston, Massachusetts, made a 3D computer model from Gage's skull—which Gage's family had bequeathed to Harlow when Gage died—showing the exact path of the tamping iron

through his head. The damage corresponds perfectly with Harlow's records.[1]

The painstaking nature of Harlow's observations are vitally important because they helped us to understand how much brain structure matters. After the accident, Gage made an excellent physical recovery but he seemed to be a different man, so much so that the same bosses who thought him a "smart and capable" and "shrewd" man before his accident refused to hire him when he reapplied for his old job as a foreman in 1849. At the time of his death, twelve years after his accident, he was working as an itinerant farmhand. One of a whole range of symptoms he experienced was a newfound compulsion to swear. Dr. Harlow wrote that Gage had become "fitful, irreverent, indulging at times in the grossest profanity (which was not previously his custom)."

For these reasons, Gage provided an important bit of evidence in the pudding wars. If the blancmange theory was correct, Gage could lose a chunk of his brain and still be left with all of his faculties intact. His brain would be the same as before—just smaller. However, Gage seemed to come out of the experience a very different person—and in a way that seemed both too profound and too specialized to be the psychological effect of having a six-kilogram missile take a shortcut through your noggin. "The equilibrium or balance, so to speak, between his intellectual faculties and his animal propensities, seems to have been destroyed," wrote Harlow. Gage's intelligence, memory, and skills were intact, but his self-control was shattered. His new habit of swearing—of using "gross profanities"—appeared to show that a vital ingredient of Gage's personality used to reside in the destroyed left frontal lobe.

!

People arguing for a differentiated brain weren't particularly close to the idea of brain structures as we understand them today. The structure-theorists of the time were, in the main, phrenologists: people who believed that you could deduce someone's personality type from the bumps on their head.

Phrenology had become popular enough as a scientific fad that articles in newspapers and magazines would frequently cite phrenological "evidence" in criminal cases. Phrenology and Gage intersected in several news articles at the time, such as an 1848 piece "Alive from the Dead, Almost" in the *North Star*, a newspaper published in Danville, Vermont. The author of the piece thought that phrenology would neatly explain the changes in Gage's condition: "Striking him on the face just below the cheekbone, [the bar] forced itself through the skull near the top of the head, passing directly through what phrenologists call the organ of *veneration*." An 1851 *American Phrenological Journal* report ("A Most Remarkable Case") went on to say that the organs of benevolence and veneration had both been damaged and that this explained "his profanity, and want of respect and kindness." This was to be the first of several times that Gage's case would be used as evidence for whatever neuroscientific theory had the upper hand over the next 150 years.

These days phrenology is rightly dismissed as a pseudoscience; the theories that phrenologists drew up were never really supported by the evidence. But those early observers of temperament and behavior at least opened the door to the idea that the brain had specialized areas. As a result, doctors began to pay close attention to brain injuries and their

consequences. The systematic study of the structure of the brain and its relation to behavior had begun and the "blanc-mange" theory would never quite regain the upper hand. As phrenology fell out of favor, scientists began to question whether the iron bar might have carried away something much more fundamental. Their answer came partially from Victorian England, and was rooted in the study of swearing.

The Victorian Origins of Neuroscience

In the late 1800s, the treatment of "lunacy"—a catchall term that included everything from epilepsy to depression, schizophrenia to the after-effects of stroke—took place in asylums. In the middle of the nineteenth century there were just 10,000 people in asylums like London's notorious Bedlam, but by the 1890s over 100,000 people were locked up, often in straitjackets or shackles, sedated with bromide, no matter what their condition.

John Hughlings Jackson was a Victorian scientist and one of the founders of neuroscience as a discipline. Born in 1835 near Harrogate to a yeoman father, he went on to study medicine in London and St. Andrews, where he was one of a new breed of young doctors who helped to develop an innovative kind of medicine rooted in deductive reasoning. It was this novel way of medical thinking that would later influence Sir Arthur Conan Doyle when he created that most famous deductive reasoner, Sherlock Holmes.

Insane asylums were open to the public as freak shows and drew large crowds. While his contemporaries visited places like Bedlam for the shock and entertainment value, Jackson took the study of "lunacy" seriously. He wanted

to know why patients suffered from certain, stereotypical types of seizures. By doing so, he made a number of significant breakthroughs in our understanding of the brain. He noted, for example, that an epileptic patient's seizures more strongly affected the side of the body opposite the site of the brain damage. This helped to prove what we now take for granted: that the left side of the brain controls the right side of the body and vice versa.

He also showed that parts of the brain work in concert with each other. For example, his observations of the way that seizures spread in the body, from the fingers or toes all the way up to the face, allowed him to deduce how the body sends movement signals via the nerves.

In his epilepsy research, Jackson also left a fascinating record of the behavior of another kind of patient—the "aphasic." Bedlam and other asylums housed many patients with brain damage that disrupted their ability to speak. Aphasics were patients who couldn't repeat what they'd said when asked, couldn't describe what they'd said and couldn't come up with new things to say. Victorian sensibilities meant that most physicians reported that these patients were completely incapable of speaking: aphasia means literally "without speech." Jackson didn't like the term "aphasia" because, he argued, it was both too specific and too general. Aphasia didn't stop his patients from speaking, like some sort of laryngitis of the mind. Rather, it robbed them of the ability to express themselves both verbally and nonverbally with gestures. But most aphasics did still speak—or at least they said words. The problem was that these words didn't seem to mean anything.

Jackson thought that the words and phrases used by

aphasics were verbal tics, woven so tightly into the patient's mind that they could no more discard them than they could stop themselves from blinking. Some of the most common of these tic-like sayings were swear words, blasphemies, and—in the case of one of his patients—only the rather sweetly vulgar "Poo!" A patient with this type of aphasia was on the receiving end of a cruel double whammy. She couldn't speak voluntarily, couldn't ask for help, or express love or longing and at the same time was stuck repeating the same few phrases over and over again in an unstoppable verbal loop.

Jackson noticed two things about these patients. Firstly, there was a difference between speaking words and understanding them. "When, from disease in the left half of the brain, speech is lost altogether, the patient understands all we say to him, at least on matters simple to him," wrote Jackson. This observation, unexceptional though it might seem, revealed something fundamentally important: speaking and understanding language are not the same thing, and are not located in the same part of the brain. Language is not a tune played by a single instrument, but a complicated orchestral score.

Secondly, Jackson noticed that brain-damaged patients could at least express *emotion* verbally, even with the few utterances that they had left, by using their tone of voice. "[The patient's] recurring utterance comes out now in one tone and now in another, according as he is vexed, glad, &c.," he wrote. "As stated already, he may swear when excited." While these aphasic patients were never going to win prizes for their oratory, they were far from speechless.

We might never have known this at all, if it weren't for the

iconoclastic Jackson's documentation of swearing among his patients. He was practically unique in this. To most observers at the time (and even some neuroscientists today) swearing didn't count as "real" speech. Instead, swearing was thought to be bestial, more like an animal's howl than human expression. Swearing was unacceptable in polite society and so it was overlooked by most of the doctors who attended Bedlam's inmates. But Dr. Jackson had no problem with recording swearing, and so as early as the 1880s swearing was recognized as a strangely resilient part of speech. Jackson's observations about the tenacity of bad language were groundbreaking, but for nearly a century after he wrote them, scientists still didn't realize quite how important that remaining profanity among aphasics is, or that the residual bad language can actually be used to communicate. We're only just beginning to understand why swearing is so good at sticking around.

There then follows an almost century-long hiatus in the study of swearing. Since there were plenty of scientists around who were prepared to study sex, death, and disease during this period, it's hard to fathom exactly why swearing was given such a wide berth. Even today I've noticed that many papers on the topic still start with an apology. I think swearing was neglected because it wasn't seen as a field important enough to overcome the "ick" factor.

This distaste or disdain for swearing meant that it went largely unexamined until the late 1980s, when researchers finally started to take the matter seriously again. Studies of brain-damaged patients in the United States, the UK, France, Germany, and Italy confirmed that swearing among aphasics is widespread, if not universal. Most of the authors

of these studies rather delicately decided not to include the swear words used by their patients, which is really frustrating; without this data it is impossible to say what sorts of swear words are most likely to be preserved.

Thankfully, this was not the case with a 1999 study by two California-based neuroscientists, Professors Diana Van Lancker and Jeffrey Cummings. They noticed that aphasic patients frequently said things like "bloody hell," "bloody hell bugger," "fuckfuckfuck, fuck off," and "oh you bugger," as well as other repetitive sayings like "well I know," "wait a minute," or "don't be sad."[2]

They studied one unfortunate patient, let's call him Charles, who had had the entire left side of his brain removed. As a result, Charles couldn't give the names of simple things that he was shown, like safety pins, measuring tapes, watches and clocks. Safety pin became "sood" and the other items he couldn't name at all. Even when asked to repeat words, he would make many mistakes, saying "November" or "sandwich" instead of "remember" and "vegent-lich" when he meant "constitution."

However, he had no problem swearing. In a five-minute recording in which he was asked to name the objects he was shown and repeat words that he heard, he said "Goddamn it!" seven times and "God!" and "shit" once. Professor Van Lancker noticed that Charles could swear far more easily than he could manage any of the other words he tried to use, and that his swearing was fluent and easy to understand while his other speech was slurred and strained. The real breakthrough came when the researchers showed a picture of Ronald Reagan to Charles. He eventually managed to name the ex-president, but initially responded with some

surprisingly fluent swearing.[3] This patient still had strong political opinions and the wherewithal to express them, with a little help from dirty language. In short, he could communicate, under the right conditions, by using swearing as a gateway to other forms of expression.

Van Lancker and Cummings wanted to pursue this line of research, but their first challenge came when they discovered that there was not, and still isn't, a standardized clinical test for swearing competence, although there is a clinical test to measure sense of humor. (More on that later.) What's more, swearing was still taboo, despite the evidence for its psychological benefits for the patient. Van Lancker and Cummings noted, somewhat bitterly: "patients are usually discouraged from using curse words during recovery as if cursing were not functionally useful during the recovery process."

Sadly, this line of research seems to have petered out through lack of support, again based on our general discomfort around swearing. This is really depressing: swearing is the one remaining means of communication that can be used by people who have suffered a massive and life-changing illness or injury, but rather than encouraging them to use what little language they have left, the emphasis is still on teaching stroke victims to control their swearing, essentially silencing what's left of these patients' voices because of our own strong feelings about strong language.

Why exactly is swearing so resilient in patients who have lost most of their speech? We don't know for sure, but swearing plays so many roles in the way we communicate —it's used to threaten, to warn, to intensify, to amuse. As

a result, swearing has deep connections in many parts of the brain, particularly to those parts that help us process emotions. And some of those parts of the brain are so fundamental that they have been with us since before we became human.

What's Right? What's Left?

There's plenty of common knowledge about the left and right sides of the brain. Most of it is bunk, or at least over-simplified to the point of near uselessness. But when it comes to swearing—and language in general—the picture is pretty clear-cut.

If you damage or lose the left side of your brain you're most likely to develop some form of aphasia. That's because some very important structures necessary for using language tend to be found here. Aphasia is a dramatically noticeable change in a person's ability to communicate and so scientists have been able to discern a lot about the workings of the left hemisphere. Although it has presented more of a challenge, scientists have also started to unpack the more subtle and hard-to-isolate effects of right-hemisphere damage.

For example, there's the case of David, who suffered massive damage to the right-hand side of his brain after a stroke.[4] David was seventy-five years old and had been bilingual his entire life. He was fluent in both Hebrew and French both before and after his stroke. According to his doctors, David was a literate and eloquent man, even after suffering brain damage, but after his stroke something strange happened to his ability to speak. You might not

notice there was anything wrong with David in the course of an ordinary conversation, but he had lost all the automatic sayings that he'd known for almost his entire life. For example, he couldn't recite the Jewish verses, prayers, and blessings he'd said every day since he was a little boy. He had lost the ability to count by rote from one to twenty, too. In fact, everything he'd learned by heart simply vanished from his mind.

These phrases and verses weren't the only thing to disappear. While David wasn't much of a swearer before his stroke, afterward he said that he never got the urge to swear at all. When researchers asked him which swear words might be appropriate in situations that were described to him he couldn't think of any, nor could he complete half-formed swearing phrases like "mother_____."

David's experience is uncommon but not unique. People with damage to the right hemisphere tend to become emotionally detached and excessively literal. These patients have problems with jokes and metaphors, can't recognize idioms, and, in most cases, completely give up on profanity even if they had been fluent swearers before. This is when a standardized test for swearing competence would come in handy.

Unlike swearing, there is a standardized clinical test for humor, called the "Humor Orientation Scales." Dr. Lee Blonder and Dr. Robin Heath applied this to other patients with damage to their right hemisphere after a stroke.[5] They discovered that these patients tend to be able to understand the logic of jokes but they just don't laugh—or even smile —when someone tells a joke around them. Also, if they try to make jokes they are usually either inappropriate or completely fall flat—perhaps because they can't imagine the likely emotional effect of the joke on other people.

Telling a joke (and understanding one) is a complex emotional process. We need to be able to model the fictional emotional states of the characters in the joke, as well as the likely emotions of the person hearing (or telling) the joke. Joke blindness hints at an emotional deficit in the brain. One of the standard tests for humor involves asking patients to complete jokes. They are read a setup and asked to pick an appropriate punchline for jokes like this:

"The neighborhood borrower approached Mr. Smith at noon on Sunday and inquired 'Say, Smith, are you using your lawnmower this afternoon?' 'Yes, I am,' Smith replied warily. Then the neighborhood borrower answered . . ."

1. "Oops!" as the rake he walked on barely missed his face;
2. "Fine, then you won't be wanting your golf clubs— I'll just borrow them";
3. "Oh well, can I borrow it when you're done, then?";
4. "The birds are always eating my grass seed."

It's not the greatest joke ever written but, for most people, option 2 is the obvious punchline. Dr. Donald Stuss of the University of Toronto found that patients with damage to the front of their right hemisphere made twelve times as many mistakes on tests like these than patients with no brain damage. Even patients who had sustained similar amounts of damage to their left hemisphere did better on the test.[6]

What's more, these patients with right hemisphere damage showed no physical responses to jokes or humorous statements. Spare a thought for the poor research assistants, forced to tell joke after joke to an audience as tough as that.

Thanks to their valiant attempts in the name of science, if not comedy, we can start to speculate on the reasons why patients with damage to their right brain might be so bad at swearing, and why ones with damage to the left brain might find it easier than any other type of speech. In most right-handed, English-speaking people at least, experiments like these show that the left hemisphere tends to house the parts that are essential for "controlled" language and the right hemisphere contains many of the parts that help us process emotion.

One of the first hints about the difference between the hemispheres emerged back in the late 1960s and early 1970s when Professor Guido Gainotti, now at the Catholic University of Rome, studied patients who had suffered damage to one side of the brain. Those whose damage was limited to the left side became very agitated, upset, and angry in response to problems they encountered during their treatment—which seems understandable, perhaps even inevitable. However, in cases with damage to the right hemisphere, Professor Gainotti noted an "indifference reaction." Nothing seemed to move these patients to even the slightest emotional response, even when they were faced with the devastating consequences of their brain damage.[7] Professor Gainotti came to what seemed like the obvious conclusion: that the right hemisphere is where emotion "resides"—those patients with left-hemisphere damage were able to respond in a natural way to their illness: with anger, frustration, and depression. Those patients with right-hemisphere brain damage were reacting "unnaturally," by not reacting at all.

That Vulcan-like detachment might seem enviable in the face of a devastating stroke, but emotion is an essential part

of our mental processing and losing it can be cognitively disastrous. Emotion works rapidly to switch our attention to things that might be a threat or a reward, and to change our behavior appropriately. In the most extreme cases it kicks in to help us flee from a threat. Emotions are quick and dirty processes that respond fast, even to confusing, fleeting, or ambiguous stimuli, before the conscious brain gets a look. In one experiment by Arne Öhman and colleagues at Uppsala University in Sweden, volunteers were shown images of snakes and spiders. The volunteers reacted with sweaty palms to images of snakes and spiders that were on screen for less than 1/300th of a second.[8] For context, it takes about half a second for the brain to categorize and identify a visual stimulus. These volunteers were experiencing responses to pictures of spiders and snakes in less than a tenth of that time, i.e., before they had actually *seen* the spider or snake. Such apparently superhuman reaction times are due to the effect of emotions on our subconscious.

This lack of access to emotional responses in patients with damage to their right hemisphere helps unpack the how and the why of swearing. Most obviously, it shows that swearing is inextricably bound up with our emotions, which suggests why patients with damage to the right brain stop swearing. One hypothesis is that these patients might no longer have the motivation to swear. If swearing is indeed the type of language that we reach for when we are angry or frustrated or joyous, then why swear if you never feel anger or frustration or joy? This is a very simple "if no emotion goes in, no swearing comes out" kind of argument.

A second and more complicated hypothesis is based on the idea that swearing is actually a very specialized and

emotionally fluent form of language that requires us to have a mental model of the emotions not just of ourselves but also of the person who hears us swearing. I'll nail my colors to the mast now and say that I think this second hypothesis seems more likely. Without the help of the right brain we can't hope to model the likely emotional response when we swear. Swearing, like joke telling, with no emotional model is like trying to navigate an unfamiliar room while blindfolded.

This second hypothesis also helps to explain why patients with left-brain damage can do little but swear. The emotions that they're experiencing are just as potent as before—possibly more so as they deal with the difficulties of living with a severe brain injury—and these emotions utilize what little speech remains as a way of making themselves heard. The remaining speech tends to comprise swear words because bad language makes use of so many parts of our brain. It doesn't just depend on the more recently evolved parts of the human brain that allow us to use swearing in deliberate, inventive ways; it also makes use of the prehistoric parts that process our emotions. This distribution of labor means that swearing is a particularly tenacious facet of our language.

Further studies suggest that we need the left and right hemispheres to coordinate their efforts when it comes to processing emotion. Your left hemisphere springs into action when you need to make *sense* of an emotion. Volunteers who were shown an emotive picture in their left visual field—which they saw in only the right hemisphere—became more emotional more quickly than when they were shown the same picture in their right visual field/left hemisphere.

The right hemisphere allowed them to experience emotions directly, like the volunteers in the snake and spider experiments. However, when the volunteers were simply asked to say whether the pictures were emotive or neutral, they were able to answer more quickly when the images were presented to the right visual field (left hemisphere). The right hemisphere is acting as a rapid alarm system—"Here's something emotional! Pay attention!"—but the left hemisphere comes in afterward to try to work out what kind of emotion we should be feeling.

This hemispheric dominance even means that the left side of your face is more expressive than the right: not only do the features on the left side of the face tend to exaggerate your expression, they're also more capable of expressing mixed emotions than the right side of the face. This isn't ideal for sending essential social signals, however: the left side of a person's face is seen in your right visual field when you're talking to them face-to-face—sending signals to the more emotionally analytical left hemisphere. This costs us processing time but allows us to make better sense of what other people are feeling.

Given that our right hemispheres do the rapid emotional processing, Dr. Tim Indersmitten and Professor Ruben Gur from the University of Pennsylvania hypothesized that we might actually make more sense of expressions on the right side of the face if these were presented for a very short period of time. They flashed up faces made of right or left sides of the face only, but because pictures of half a face look so strange, Indersmitten and Gur did something very cunning. They used "symmetrical chimeric faces"—photographs of faces where either the left or right half is duplicated and

Left-left composite   Right-right composite

Figure 2: Symmetrical chimeric faces

flipped to make a symmetrical face (Figure 2). Although the facial expressions were indeed more exaggerated in pictures made up of two left halves of the face, volunteers made far fewer mistakes when interpreting the emotions in the pictures made up of two right halves of the face, particularly when asked to make decisions in under six seconds.[9]

If you stare at the photographs above for more than a few seconds, you're likely to see the exaggerated expressions on the left as being much clearer. But in real life we have very little time to make sense of facial expressions with our left hemisphere—they tend to be fleeting and last a matter of

a few seconds at most. It turns out that rapid glances at the right side of someone's face (in our left visual field) go straight to the right side of our brains and allow us to make quicker, more accurate judgments about someone's emotional state.

But the lateralization of the brain—the division of labor between left and right hemispheres—isn't the whole story when it comes to emotions and swearing.

Introducing the Amygdala

The fact that there is a left and right brain—and that one is responsible for reason and the other for emotion—might seem very familiar. It's one of those folk neuroscience stories that turn up everywhere from self-help books to management training seminars. But, as we've just seen, for most of us the two sides of the brain work in concert to help us deal with both emotion *and* reason. And they don't do it alone: there are other parts of the brain that have specific roles in either provoking or controlling the emotions that lead to swearing.

Meet your amygdalae: an amygdala is a small, almond-shaped (and roughly almond-sized) node of the brain and you have one on each side. If you imagined a line going straight from ear to ear and another through each eye (try not to think too hard about Phineas Gage) the points where they intersect are where your amygdalae can be found.

When we talked about the differences between the hemispheres, we were talking about developments that are quite recent in evolutionary terms. Some of the complex structures

and processes we have in the cortex are unique to primates. Some language-specific areas are only found in humans. The amygdala, on the other hand, is found in all mammals, and there are similar structures in the brains of reptiles, fish, and birds. This means that the first, rudimentary amygdalae probably appeared around 250 million years ago, which is the last time we shared a common ancestor with the pigeon, the sturgeon, and the frog.

Why are we carrying around such an ancient piece of hardware when we have such wonderfully sophisticated structures in the cortex? The answer is that those sophisticated structures couldn't function without signals from these ancient little bumps located deep behind your ears. For something so small and so simple, the amygdala has a lot to do. We know that people with bigger amygdalae are better at making and keeping friends, for example, and amygdala size is a good indicator of whether or not you will suffer from depression. Our amygdalae are busy emotional relays that let the rest of our brains know when we're fearful, anxious, or sexually aroused.

The brain itself has no pain receptors, so if you're unlucky enough to need brain surgery it will probably be carried out with local anesthetic for your scalp while you are wide awake. Far from being barbaric, keeping a person conscious during an operation has three advantages: first, general anesthetic occasionally kills people; it's better avoided if possible. More important, the surgeon can use a small electrical current to send signals to parts of the brain before cutting. That might sound terrifying, but because the anatomy of our individual brains is as varied as the rest of our bodies, mapping out your particular brain's anatomy is essential to ensure that the surgical

team aren't about to remove a vital part of it. Finally—and very usefully for science—observations of responses during surgery can provide information about the function of a living brain that we would otherwise have no way of gathering. This is why we know for certain what it's like to speak to people whose amygdalae are being artificially stimulated.

Two Scottish doctors—Edward Hitchcock and Valerie Cairns—had this conversation with a thirty-four-year-old man undergoing surgery:

Hitchcock (who is operating on the patient): "How are you now?"

Patient: "Just the same."

Then the patient's amygdala is stimulated electronically.

Patient: "I can hardly speak. **** . . . I just want to get the **** out of here." [The swearing is redacted in the original paper so sadly I can't share with you exactly what he said, though I can make a good guess.]

Hitchcock: "That's OK." (Turns off stimulation.) "All right?"

Patient: "Yes."

Hitchcock: "Did you feel angry?"

Patient (surprised): "Aye, I did."

Hitchcock: "Do you feel that now?"

Patient: "No, I don't feel that now."

The noteworthy thing here is not so much that the patient swore while under the surgeon's knife—a more rational response is hard to imagine!—but that his outburst was so sudden, short-lived and surprising even to him.[10] We already know that the amygdala's major function is to tell us when we are feeling especially emotional—and so the fact that its stimulation provoked swearing suggests not

only that the amygdala is essential to the production of this type of language, but also that our emotions are inextricably bound up with our impulse to swear.

From surgeries like these, scientists now know that removal of the amygdala reduces emotional responsiveness in general and aggression in particular. As a result, it is thought that the amygdala has a role to play in suppressing unwanted swearing, by acting as a sort of emotional traffic light and letting us know how and when it's appropriate to express anger or fear. Signals from the amygdala can encourage us to let rip with bad language when we want to show anger but don't fear reprisals, while different signals from the same small node can warn us when doing so might be dangerous. For such a small and primitive part of the brain, it's capable of some complex processing.

Swearing Is a Team Effort

In the years since Phineas Gage's accident we've learned that different parts of the brain do indeed have different functions—but we're also starting to learn that none of them works entirely in isolation. So many parts of your brain are involved in swearing; either collaborating to help you produce swearing or working to suppress it when it isn't wanted. The brain, unsurprisingly, is neither a trifle nor a blancmange. It is, though, a bit like an orchestra; a set of highly specialized elements working together to create what seems for all the world to be a unified whole.

Swearing relies on sophisticated structures in both right and left hemispheres but it also draws on one of the most

primitive parts of the early brain. What does this mean in practice? Well, if swearing were a simple, primal thing we wouldn't expect to see so much involvement from the more recent, more complicated areas of the brain. However, if swearing weren't so closely linked to our emotions, the amygdala wouldn't have such an important role to play. And the fact that swearing is lost if we lose our ability to map the emotions of others shows us just how socially sophisticated we have to be in order to swear.

It might seem like a contradiction—swearing is both primitive and sophisticated—but, as later chapters will show, this argument makes sense in the context of how humans evolved. We developed language as a means of getting along socially: the primates who told each other about the tiger they'd seen in the trees were much more likely to survive than the ones who didn't. We also learned to communicate complex emotions like "I'm angry, back off!" and "I'm stressed, give me what I want!" Swearing is a powerful shortcut—an emotionally freighted part of language that lets us communicate complex things in an urgent way. What might have started out as simple displays of fear and aggression have become more elaborate as our societies, and our brains, have grown and changed. It's no wonder, then, that when it comes to the brain, swearing is a sophisticated team effort.

*₂ * * * * *

"Fuck! That hurts."
Pain and Swearing

Dr. Richard Stephens, psychologist and author of *Black Sheep: The Hidden Benefits of Being Bad*, is infectiously enthusiastic about swearing. Every year he and his undergraduate students come up with more exciting experiments that are helping to untangle the relationship between pain, emotion, and swearing. "Swearing is a great vehicle for teaching psychology. Everyone's fascinated because everyone does it, but the experiments also show how important it is to have a control condition, that it's important to have an evidence base, and the ways that logic and science work." Dr. Stephens talks about his undergraduates at Keele University in Staffordshire as "a fantastic crop of students." Surely he's the university lecturer any aspiring psychology undergraduate would want? That rather depends on your attitude to sacrificing comfort for science: Dr. Stephens makes his students suffer for their results.

For a very long time, conventional wisdom has said that swearing is not a useful response to pain. In fact, many

psychologists believed that swearing would actually make pain feel worse thanks to a cognitive distortion known as catastrophizing. Cognitive distortions are exaggerated ways of thinking that make it hard to cope or to act rationally. When we catastrophize we leap to the conclusion that the bad thing that is currently happening is the absolute *worst* thing. We're usually catastrophizing when we say things like, "This is terrible! I just can't!" Swearing was thought to reinforce that feeling of helplessness.

If swearing is part of a catastrophizing response, it would indeed make us less likely to cope well. Thinking "I just can't!" doesn't usually help us to be resilient in the face of pain and adversity. But this troubled Dr. Stephens, who wondered "why swearing, a supposedly maladaptive response to pain, is such a common pain response." Like all of us, he's hit his thumb with a hammer enough times to know that swearing seems to be an unavoidable response. So, with his students, he set out to find out whether swearing really does make pain feel worse.

The first of these studies resulted in the paper that inspired this book. Somehow, he persuaded sixty-seven of his undergraduate students to stick their hands in ice-cold water for as long as they could stand, and do it not just once but twice, once while swearing and once not. (The Keele University School of Psychology Research Ethics Committee approved the study, which might be something to ponder if you're choosing your future alma mater.) The thinking behind the experiment was as follows: if swearing is so maladaptive then the volunteers would give up much faster while they were cursing than if they were saying another, neutral word.

To make it a fair test, the students were allowed only one swear word and one neutral word and the order of the swearing and neutral immersions was randomized. Dr. Stephens asked them for five words they would use if they dropped a hammer on their thumb and five words to describe a table. Then he took the first swear word that appeared in the first list and its counterpart from the second list. When I did the experiment, my words were: "arrgh, no, fuck, bugger, shit" and "flat, wooden, sturdy, shiny, useful" which meant saying "fuck" in one trial and "sturdy" in the other.

The results could best be summarized by the phrase "Maladaptive, my ass!" It turned out that, when they were swearing, the intrepid volunteers could keep their hands in the water nearly half as long again as when they used their table-based adjectives. Not only that, while they were swearing the volunteers' heart rates went up and their *perception* of pain went down: in other words, the volunteers experienced less pain while swearing.[1]

It's an easy experiment to try for yourself at home, or at a party if you have the right kind of friends. All you need is a bowl of ice water and a stopwatch. So why wasn't this experiment done soon after the invention of the ice cube?

"Pain used to be thought of as a purely biological phenomenon, but actually pain is very much psychological. The same level of injury will hurt more or less in different circumstances," says Dr. Stephens. We know, for example, that if male volunteers are asked to rate how painful a stimulus is, most of them will say it hurts less if the person collecting the data is a woman.[2] Pain isn't a simple relationship between the intensity of a stimulus and the severity of your response. Circumstances, your personality, your mood, even

the experience of previous pain all affect the way we experience a physical hurt.

Now that we know how much influence our thoughts, feelings, and experiences have on our perception of pain, psychologists have started to look at ways of influencing our brains to make pain easier to tolerate. Dr. Stephens's study shows that swearing is affecting the volunteers' perception of pain, but how? Is it a simple distraction? Does swearing make us feel stronger? Does it simply let us vent some of our feelings? To find out how it helps, Dr. Stephens and other researchers have started to look at the phenomenon in detail. What we're learning about the link between pain and the emotions, and between our bodies and our words, is fascinating.

What Does Swearing Do to the Brain?

One of the most exciting things about the original ice-water study is that people don't just say that pain is less painful when they swear, nor do they simply experience it as easier to bear. Swearing seems to change something in their bodies. As noted, Dr. Stephens's study showed that the volunteers' heart rates went up while swearing, compared with the non-swearing trials. An increase in heart rate is a sure sign of our emotions being involved.

In psychology experiments, showing people swear words, or having them listen to some swearing, has been used for a long time as a way of making people experience strong feelings, but it's only in the last eight years or so that psychologists have started to study the effect of swearing on

the volunteer by having them do the swearing themselves. Dr. Stephens thought that the likely effect on our emotions made swearing an excellent candidate for an analgesic. He has a theory that swearing does something to the emotions that makes pain easier to stand and that the two most likely candidates are fear and aggression.

We know that fear can act as a painkiller. Dr. Jamie Rhudy and his colleagues at the University of Tulsa, Oklahoma, have been studying the effect of emotion on pain for several years. In one experiment the volunteers were told that they would, might, or would not receive mild electric shocks during the experiment. (I don't know about you, but if I volunteered for a psychology experiment and the first thing the experimenter said was, "You will not receive any electrical stimulations," I'd be deeply suspicious.)

None of the volunteers got off lightly, though. All three groups were asked to keep one finger close to a "radiant heat source" (a high-powered projector bulb) for about 20 percent longer than was comfortable. The volunteers who had simply been threatened with the shocks reported that the exposure to the heat source felt less painful than those who had been told they wouldn't be shocked.[3] What's more, the volunteers who actually were shocked reported that the pain from the bulb was even less intense than those who'd merely had the prospect of electric shocks hanging over them.

What could be going on here? Might the shocks have numbed the pain? Dr. Rhudy had the shocked group carry out the heat tolerance test with the unshocked hand in order to rule out any localized numbing effect from the electrical current. The results were the same. Indeed, as we'll see later in this chapter, although pain can have an overall numbing effect, it has to be of limb-breaking severity to do so.[4] Dr.

Rhudy concluded that the anticipation of pain—a state of fear—is enough to make us preempt the sensation and reduce its apparent strength. In short, the more pain we fear, the less pain we feel.

However, this runs counter to the numerous other studies that suggest that anticipation and fear actually increase your subjective experience of pain. Because it's not ethical to ask people to volunteer to be tortured, there aren't many experiments about the anticipation and experience of extreme pain. Fortunately for the research community, there is one group of people who are pretty much guaranteed to be in a significant amount of pain at some approximately known date in the future, many of them choosing to accept little or no pain relief. Pregnant women make excellent study candidates.

From them, we now know that one of the strongest predictors of how painful a woman will find labor and childbirth is the extent to which they are afraid that it will hurt.[5] Fear of childbirth can also make women less resilient to pain that is unrelated to the birth. In the ice-water experiment, women who are anticipating a painful delivery tend to remove their hands from ice water faster than those who are confident that everything will go well.[6] That's why, at the time of writing, I'm actively ignoring horror stories about childbirth, to the extent of putting my hands over my ears and singing, "Lalalala I can't hear you." With just a few weeks to go until the birth of my first child I'm not being rude—it's evidence-based pain management.*

---

* I managed twenty-two hours of labor with gas and air and no swearing. On this highly unscientific basis, I conclude that the secret to a manageable labor is a long, thin baby with a narrow head!

According to the experience of pregnant women at least, then, fear makes pain more painful, as should priming with a mild pain, which appeared to contradict Dr. Rhudy's results. Dr. Rhudy revisited his study and, while he didn't go to the extent of recalling his original volunteers, he did pay more attention in future research to the range of emotions caused by both pain and the threat of it. One conclusion he reached was: "Our laboratory has found that procedures intended to elicit fear also elicit anger."[7] So, shocking people also makes them angry? That's definitely something to keep in mind when designing an experiment that involves electricity.

I'm being a little facetious, I know, but this result highlights one of the dangers of trying to study emotional responses. Emotions come not as single spies but in battalions. Inducing one pure emotion in a person is impossible. How can we study the emotional effect of swearing when emotions are so difficult to unravel? Psychologists classify emotions along two axes: valence and arousal. Valence simply refers to how pleasurable (or not) a feeling is. Happiness scores highly for valence, misery has low valence. Arousal is a measure of how strong a feeling is, regardless of whether or not it is pleasant. So excitement and fury are both high arousal, while boredom and contentment are both low.

When studying the effect of swearing, Dr. Stephens doesn't assume that swearing has induced a particular emotional state in all of his volunteers. Instead he, like many other psychologists, quantifies the degree of each volunteer's arousal using their heart rate and galvanic skin response (roughly speaking, a measure of how sweaty-palmed you are; researchers attach small electrodes to

volunteers' fingertips. These detect levels of stress, fear, anxiety, or excitement).

In the first of the ice-water experiments, Dr. Stephens showed that swearing really did change the volunteers' arousal levels. "As well as making the ice water *feel* less painful, we also showed that swearing causes effects on various parts of the body. It does increase heart rate: it seems to cause the fight-or-flight response. So if we think that swearing can help with pain because it causes emotional arousal, then what about doing something that just causes emotional arousal?"

Dr. Stephens designed a particularly cunning experiment with one of his undergraduates, Claire Allsop.[8] This study was so neatly devised that she won a prestigious award from the British Psychological Society for it. Ms. Allsop wanted to know whether she could increase pain tolerance by making someone feel more aggressive. If pain tolerance depends on "innate" aggression then it shouldn't be possible to induce mild-mannered people to suffer for longer. But if, as the swearing study showed, the same person can stand far greater levels of pain when swearing than when not, might swearing actually cause aggression levels to rise, increase arousal, and help us deal with pain that way?

She followed in her mentor's footsteps, and managed to persuade forty of her fellow undergraduates to repeat the ice-water test.* "We were looking at things we could do in the lab and one easy way is to have them play a first-person shooter game," explains Dr. Stephens. In fact, each of her volunteers played either a first-person shooter—one of those video

---

* I can only assume that Ms. Allsop was very popular among her classmates. Before the study, at least.

games where you run around trying to kill people before they kill you—or a golf game. In order to test exactly *how* the game had affected the volunteers, Miss Allsop then had them fill in a hostility questionnaire where they rated themselves from 1 to 5 against adjectives like explosive, irritable, calm, or kindly. Finally, she used a very clever test to see how aggressively primed the students were. The test is a kind of solitary hangman—she showed the volunteers prompts like "explo_e" or "_ight." Those who responded with "explode" and "fight" she classified as feeling more aggressive than those who thought of "explore" or "light."

The students scored consistently higher on the aggression measures when they played the shoot-'em-up rather than the golf game, rating themselves as more hostile on the questionnaire and coming up with more violent imagery in the solo hangman challenge. But did it do anything for their pain?

"We basically showed the same pattern of effect as we did for swearing: they could tolerate [the ice water] longer, and said they perceived it as less painful, and they also showed a rise in heart rate." After the golf game the male students could immerse their hands for an average of 117 seconds, females an average of 106 seconds. After shooting people, those times jumped to 195 seconds for the men and 174 seconds for the women. That's around *three minutes*. If you're in any doubt whether or not that's a noble feat I defy you to try it. We did the same experiment in our laboratory (somewhat informally), comparing swearing with positive affirmations like, "Emma, you can do it." I couldn't. I've lost my notes, but I think I lasted all of ninety seconds—much shorter than my swearing best, which was just over three minutes.

Does this mean that people who are inherently aggressive are more likely to handle pain better? To test this, as part of her undergraduate research Dr. Kristin Neil and her colleagues at the University of Georgia looked at the relationship between how aggressive someone is and how much pain they can stand. She asked seventy-four male undergraduates to take part in a set of "reaction-time contests," ostensibly because she wanted to check how fast the students could press a button. But the real reason was rather different.

In Dr. Neil's lab, volunteers were given "reaction buttons" to press. They were told to imagine themselves like gunslingers in a western—they had to be faster than their (unseen) opponent at pressing the button after a cue in order to win the game. She also introduced an interesting wrinkle. Next to the reaction button was a punishment button. If their opponent was thought to be cheating, or even if the volunteer was getting frustrated at losing and wanted to even up the odds, the punishment button would administer an electric shock for as long as it was pressed. The intensity of the shock could be decided by the volunteer. In order to give the volunteers some idea of just how much punishment they would be meting out, Dr. Neil gave them all a series of shocks before the game began, increasing the level until the volunteers asked her to stop.

All was not as it seemed, however. The opponent in the game was nothing more than a simple script on a computer that would let the volunteer win a certain percentage of "gunfights." The punishment button merely recorded the intensity level and how soon, how often, and how long

the volunteer pressed it. Of course, the real experiment had begun long before the game started. With those initial shocks, Dr. Neil was covertly collecting data to see how much pain each volunteer could tolerate.

All this deception might seem unethical, but it's common practice when designing psychology experiments. Humans are social animals; we are strongly conditioned to be cooperative and to behave in ways we think are expected of us. If a volunteer expects to be tested on something, they will do their best to behave in the way that they think an ideal subject would behave, even going so far as to underperform if that's what they think the experimenter wants to see. This subconscious desire to please can throw results off very badly, so Dr. Neil had to counter little white lies with bigger white lies. While her volunteers were distracted by the idea of doing a good job on the reaction-time experiment she could collect the data she really needed.

What she wanted to know was whether there is a correlation between a person's pain threshold and how soon, how hard, and how often they punish their opponents. The results are indisputable: the more pain a volunteer was able to take before the trial, the more likely they were to shock sooner, more often, at higher voltage and even to lean on the button for longer than their less pain-tolerant fellows.

Why should that be the case? Do the less pain-tolerant volunteers have greater empathy for their "victim," or is there something about the most aggressive players' brains that allows them to suck up more discomfort? Dr. Neil's experiment doesn't look at this directly, but by comparing the

results she got with the results that Clare Allsop and Richard Stephens uncovered, we can build some hypotheses.

We know that our level of aggression at any given moment is a combination of the aggressive elements of our personality (known as trait aggression) and our reaction to present circumstances (state aggression). Dr. Neil's experiment *seems* to suggest that individuals with high trait aggression are better at withstanding pain, but the more aggressive volunteers might also have been having very bad days: the experiment doesn't disentangle state and trait aggression explicitly. What's so great about the Allsop and Stephens study is that it shows how easily we can all manipulate our emotions as a means of managing pain. Does that mean that swearing—or shoot-'em-ups—should be available on prescription?

Is All Swearing Equally Good at Killing Pain?

The good news is that swearing and shoot-'em-ups seem to work for everyone that Dr. Stephens has studied. Psychologists classify people into those who tend to express their anger a lot ("anger-out" people) and those who sit on it ("anger-in" people.) At first Dr. Stephens suspected that swearing might only work for people who were comfortable with the idea of swearing, or who did a lot of swearing in their everyday lives. "You might think that swearing would be more effective in anger-out people than anger-in people because for anger-out people you're giving them an outlet they're used to using whereas for anger-in people it's maybe the opposite of what they might want to do." He set up an experiment to test this, asking people to rate how likely

they were to swear when they were angry, but the results surprised him: "Actually it didn't make a difference; swearing worked equally well for both types of people. That's the thing about science: sometimes you get a negative result."

There are other variations that Dr. Stephens and his students have tested. "When the first paper came out, the question that most people seemed to ask was, 'Does it matter how much I swear in everyday life?'" The first study that looked at this phenomenon seemed to suggest that there was an effect: swearing helped all the volunteers, but the extra tolerance time it bought the frequent swearers was a lot more limited than the relief that was felt by infrequent swearers.[9] But the key to scientific discovery is repeatability, and Dr. Stephens says that follow-up studies haven't replicated this effect.

The type of swearing might make a difference, though. What about "minced oaths"—those socially palatable versions of swearing we trot out when we might be overheard? Do these milder types of naughty language work as well when we want to get our aggression rates up? It seems not: stronger swear words are stronger painkillers.

"My students tried to see if there was a dose response for swearing," says Dr. Stephens. Two students ran a variant of the same experiment in two consecutive years that looked at the relationship between the strength of the language and the effect on pain. One year a student compared saying "fuck," "bum," or a neutral word. The following year another student did the same experiment but thought that "bum" was too mild and so decided to use "shit" instead. In both experiments, "fuck" gave the greatest relief, while "bum" and "shit" gave less, though more than using a neutral word.

While the study was a classroom-based curiosity that hasn't been published, it does sound like a promising avenue for further research, as well as making for an entertaining talk: "I love putting that slide up in presentations because I get to say the word 'bum,' which is quite fun."

The result also suggests a converse experiment to me: can we rate the severity of swear words by how much analgesic effect they have? Rather than asking people to say subjectively whether they think a swear word is mild, moderate, or severe, wire them up to heart rate monitors and have them stick their hands in ice water. Perhaps that's something for the team at the *Oxford English Dictionary* to consider ahead of their next edition.

Social Pain Is Real Pain—and Swearing Helps There Too

The relationship between physical pain and emotional states is definitely a complicated one, made more so by the fact that we all experience something called social pain. Social pain, feelings of being rejected or excluded, is as real as physical pain. Experiments with acetaminophen[10] and marijuana[11] (not at the same time) show that identical analgesics can relieve both social and physical pain. It makes a lot of evolutionary sense. For most of human history, experiencing loss or rejection could have been as detrimental to your survival as appendicitis or a broken leg.

One of the most astounding experiments to demonstrate the equivalence between social and physical pain looks at the way two pains that are experienced in quick succession tend

to interact. We know, from other studies, that two physical pains experienced in quick succession have an entirely unexpected effect on the way we perceive them. A mild pain makes us temporarily more sensitive to discomfort whereas severe pain numbs us and makes us more able to bear further trauma.[12] There might be an excellent reason for this: if you're bitten by a dog, the fight-or-flight instinct kicks in. We become highly vigilant to other pains either as extra motivation to get out or fight back, or as a way of avoiding further trauma in our fight or flight. In contrast, for the kind of pain where curling up in a defensive ball is the best course of action—a broken limb, for example—further pain tends to feel much less severe than it would otherwise. We can stand a further mauling, because fighting or fleeing are not an option.

If social pain is activating the same psychological or biological pathways as physical pain then moderate social pain should make further physical pain seem more intense whereas severe social pain should numb us to physical trauma. We now know that this is actually the case thanks to one of the most disturbing of all pain-tolerance studies.

Volunteers were randomized to one of two types of social pain.[13] In the first group, people were asked to play an online game of throw and catch. They weren't told that they were actually playing against computers, so when the ball was never thrown to them they felt left out and shunned. That was the mild version of social pain. Members of the other group were asked to fill in a questionnaire about their personality and preferences and were told that their answers would be assessed by a highly reliable software model that

could predict the strength of their future relationships. However, the researchers completely ignored the answers and gave every volunteer in the "severe social pain" category the same profile:

> You're the type who will end up alone later in life. You may have friends and relationships now, but by your mid twenties most of these will have drifted away. You may even marry or have several marriages, but these are likely to be short-lived and not continue into your thirties . . . The odds are you'll end up being alone more and more.

It's not often that "and then we pinch them really hard" is the *least* upsetting part of one of these experiments. But when the volunteers were subjected to a number of carefully calibrated pinches afterward, those who missed out on the ball in the game of throw and catch found them much more painful than the people who were told that they were unappealing losers, destined for a life of loneliness and misery.

And, of course, if social pain is the same as physical pain we should find that swearing reduces its severity, too. And that's exactly what Laura Lombardo and Dr. Michael Philipp of the University of Queensland, Australia, found. They asked volunteers to remember an experience of being excluded from a group or included in a group. They found that swearing after being made to recall an event where you felt hurt reduces the pain associated with the memory of that hurt.[14] But swearing can also increase the likelihood of being excluded, especially if you're a woman.

Swearing and Illness

Cancer is such a social taboo in some countries that the name itself is even a swear word.* It's probably not surprising that people going through long-term illness have a tendency to swear. Pain, fear, and frustration are all good triggers for bad language and a little bit of swearing can help us bear up under the pressure.

Dr. Sarah Seymour-Smith at the University of Huddersfield studied the experiences of men who were suffering from testicular cancer. They found going to support groups to talk about feelings extremely difficult,[15] but they did find swearing really cathartic. She includes a transcript from the video diary of "Cal," who has just had an operation to have a testicle removed.[16] He's listening to "Always Look on the Bright Side" while he makes a video in private, which he says that he doesn't intend to show anyone. He starts by talking about how "shit" the year has been, and says that it's "really bollocks, er . . . bollock sorry!" Cal is definitely reluctant to go to a self-help group to talk about his feelings, anything where he might have to "stand up, you know, 'my name's Cal Jackson I've had testicular cancer' and then burst into tears and all that sort of thing and blokes don't like that sort of thing. I don't want anybody to cry in front of me or anything like that." But swearing on his video and, most likely, with his mates, is giving him an outlet to deal with his feelings.

"I think Cal is adopting a masculine image in his video diary as a way of making light of the issue," says Dr. Seymour-Smith. "Humor is something that men do seem to do when

---

* "Kanker" in Dutch, for example.

they're ill, probably as bravado. When I've studied testicular cancer support forums humor is evident there." The men on these forums tend to use words like "womble" (which sounds like "one-ball") or speak of being a member of the "flatbaggers club." Dr. Seymour-Smith's research suggests that talking in this taboo-breaking, jokey manner is "a way of reworking a positive identity from having both testicles removed." A means of both dealing with the pain and reasserting his masculinity.

It's far more socially acceptable for women to talk about their feelings, even to cry, but if they vent their emotions with swearing it doesn't go well for them. In a lab, with a bucket of ice, swearing helps women as much as men, but in the real world, with long-term, life-changing pain, women lose out when they swear.

Professor Megan Robbins and her colleagues at the University of Arizona wanted to know whether women with breast cancer and other long-term conditions swore, and, if so, whether it did them any good. From everything that we know about pain and swearing you might expect that a good swear would help these women cope better with their illness but, in a finding that both surprises and depresses me, women who swore ended up more depressed and had less support from their friends than those who were less likely to let rip with the swear words.[17]

To carry out her study, Professor Robbins recruited a group of women volunteers who had recently been diagnosed with breast cancer. She gave them all voice-activated tape recorders that caught about 10 percent of their speech over a weekend. She analyzed that data and also followed up with these women to see how they were getting along sometime later. Sadly, she found that the women who swore more

around their female friends tended to lose those friends over the course of their illness and to end up more depressed.

Comparing these women's experience with the experiences of Cal and his friends reveals something quite depressing. Women, certainly women born before 1960, are still uncomfortable around swearing. "I think it's a combination of age and gender," says Dr. Robbins. "The women in my paper were in their late fifties. In that generation they weren't socialized to swear quite like men do." I ask her if she thinks this will change. As a frequent swearer herself, she hopes so: "I can certainly identify with the [people like Cal]—sometimes you swear with your in-group and it promotes bonding, but not so in my parents' generation."

It's not that these women were being objectionable, swearing angrily at their friends and alienating them. "Because we saw this decrease in emotional support I thought, 'Oh my gosh, they're swearing *at* people.' But we found that that almost never happened. It was also rare that they were swearing in the context of their illness." Most of the swearing was "kind of casual. Like the patient who was working with her husband on a project and she said to him, 'So how do we bust up that shit?' It's this harmless swearing that was happening but even still the women who swear the most tended to fare worst." As we'll see in chapter 6, some of society's attitudes to women and swearing are still extremely reactionary. Big boys don't cry and nice girls don't swear, even in the twenty-first century.

As Dr. Robbins says:

If my generation were in this study down the line I hope it would be different. As somebody who swears—to be

funny, or for bonding—I think it is unfortunate for these women who are in midlife and are, you know, women. It's important that our friends know there are circumstances where swearing might help this person who's coping with something really painful. If you hear swearing, don't go away.

There's no doubt that it can be hard to hear a torrent of swearing—after all, we've seen now how emotive swearing can be. But from what we know about swearing and pain, and from examples like Cal's experience with losing a testicle to cancer, swearing is a really important part of dealing with the shitty consequences of pain and illness. If we drop out when our friends sound off, are we really being friendly at all?

* * ₃ * * * *

Tourette's Syndrome, or
Why This Chapter Shouldn't
Be in This Book

When I tell people that I'm writing a book about swearing, one of the first things they usually ask about is Tourette's Syndrome (TS). My answer usually disappoints them: for all its colloquial associations with swearing, 75 percent of sufferers don't experience compulsive bad language at all. The second part of my answer is usually even more unsatisfying: even in those patients who do swear, no one is entirely sure what causes TS. We can make certain inferences. For example, the fact that certain drugs alleviate tics suggests that the neurotransmitter dopamine is somehow involved, but we still aren't entirely certain how. We also know, from family studies, that TS is a genetic condition, but we don't yet know which genes are involved or why.

We also know that most TS sufferers are plagued with a host of other problems such as obsessive compulsive disorder (OCD) and attention deficit hyperactivity disorder (ADHD). Not only does it make these people's lives even more challenging, it also makes it hard to pin down

the definition of where TS ends and some of these other illnesses begin.

For a start, while TS might be popularly associated with swearing, in reality it presents as a much more varied set of symptoms. According to Professor Timothy Jay, a psychologist at the Massachusetts College of Liberal Arts, fewer than a quarter of TS sufferers have problems with uncontrollable swearing.[1] Even that rather imprecise number is up for debate. Depending on how you define coprolalia (*copro*, Greek for "shit" and *lalia*, Greek for "chatter"—talking shit has a medical name), then anywhere between 7 and 40 percent of TS patients suffer from the urge to use bad language.

One thing is clear, all TS sufferers are afflicted by "unvoluntary"* movements: it's part of the definition of the illness. According to the Centers for Disease Control and Prevention, someone suffering from TS will:

- have two or more motor tics (for example, blinking or shrugging the shoulders) *and* at least one vocal tic (for example, humming, clearing the throat, or yelling out a word or phrase), although they might not always happen at the same time.
- have had tics for at least a year. The tics can occur many times a day (usually in bouts) nearly every day, or off and on.
- have tics that begin before he or she is eighteen years of age.
- have symptoms that are not due to taking medicine

---

* Unvoluntary movements, as distinct from involuntary ones, serve no intrinsically useful purpose. A blink is involuntary; compulsively making the wanker gesture in one's workplace is (usually) unvoluntary.

or other drugs or due to having another medical condition (for example, seizures, Huntington disease, or postviral encephalitis).[2]

These tics aren't like the involuntary blinks or coughs that the rest of us experience. Instead they are movements that the person doesn't want to make, and can even sometimes suppress, but that generally get harder to control when they are stressed or tired. Sufferers tend to experience bouts of urges, where a number of tics occur in clusters and the severity of these urges comes and goes. For most sufferers, the urges of TS tend to wear off, or at least diminish, sometime in late adolescence.[3] While around six in every thousand school-age children suffer from TS, numbers drop to around one in two thousand for adults.[4]

For almost all patients, their tics tend to develop in a fairly stereotypical pattern. According to Dr. Christine Conelea, assistant professor of psychiatry and human behavior at Brown University, patients usually start to get the urge to make movements of the eyes, face, or head, beginning as blinks and sniffs. After a while, these motor tics start to spread down the body, becoming more complex as they go. For some children, what starts as a small shoulder spasm might develop into an elaborate arm movement over a period of months or years.

The same sort of thing happens with vocal tics, which tend to start as compulsive sniffs, throat-clearing, or simple noises but develop into whole words or phrases.[5] But not all of these will result in swearing. In fact, while many people jokingly refer to any frequent swearing as "Tourette's" there are other illnesses that are much more likely to result in

bouts of bad language. We know that schizophrenia and some personality disorders are strongly linked with swearing, which is usually "propositional"—it's planned to have a specific effect on the hearer—whereas TS sufferers have no such control over their vocal tics.

In the first chapter we also saw how some stroke patients with aphasia are able to swear, despite not being able to say much else. But there are some marked differences here, too. Stroke patients can learn to use swearing in a propositional way but their swearing vocabulary is limited to those few swear words that are committed to rote memory before the stroke. In contrast, TS sufferers can (and often do) learn new swear words or phrases that become part of their vocal tics over time.

The coprolalia in TS is different from these types of illness-related swearing, and can be described as "unpropositional." The patient can't control their outbursts—or at least not easily—but they aren't provoked by pain or anger. In a distressing irony, for most TS patients, their swearing is provoked largely by how socially unacceptable it would be and, as with swearing for pain relief, the stronger the swear word, the greater the satisfaction. But the choice of "forbidden" words isn't necessarily restricted to swearing. At least one patient suffered from the distressing urge to shout their ex-lover's name in front of their current partner, causing extreme awkwardness all round.[6]

Coprolalia in TS seems almost to be designed to cause the sufferer maximum stress and embarrassment. The swear words and other inappropriate language that comes out during bouts of coprolalia tend to be louder and clearer than the rest of what that person is saying, which draws attention

to the tic and increases their mortification. While some sufferers can suppress the impulse to use bad language, it tends to cause them a great deal of anxiety. Remember the immense irritation of trying not to sneeze or scratch an itch. Now imagine that giving in to the sneeze or the scratch could get you anything from a dirty look to being arrested. Some TS sufferers try using "minced oaths" like "sugar" for shit or "fudge" for fuck. But, as with pain relief, these disguised forms of swearing give much less relief, like rubbing that itch instead of scratching it.

Frustratingly, it's hard to study how coprolalia develops because we don't know exactly which swear words or other inappropriate outbursts plague TS sufferers. Professor Jay has made an extensive survey of the research into TS and he laments that so many researchers in this field are exasperatingly coy about the exact language that TS patients use. "Many authors who study Tourette's syndrome unaccountably don't give the specific swear words that are used by the coprolaliac patients. Some refer only to 'uncontrollable strings of obscenities,' or 'a number of four letter words.'" However, in the data that he has personally managed to collect, the top five swear words in coprolalia among TS sufferers who speak English are, in descending order: "fuck," "shit," "cunt," "motherfucker," and "prick."

Like all swearing, the use of bad language in TS is extremely culturally specific. According to Professor Jay, in Japanese culture, rather than swearing, TS sufferers are more likely to feel compelled to hurl insults such as "lecherous," "stupid," or "ugly." In other languages patients tend to use commands like "shut up" (*kaeft*) in Danish, or "shut up, stupid" (*taci, cretinaccio*) in Italian; while in German, TS

patients tend to use vivid, non-swearing taboos like "rotten bones" (*verfaulte Knopfen*).

The embarrassing outbursts experienced by TS sufferers aren't confined to swearing, insults, or ex-lovers' names. Those urges also include coprographia (the urge to write swear words) and copropraxis, which usually consists of rubbing or touching your genitals in public. Copropraxis is that set of gestures that is rude but doesn't actually mean anything (other than the risk of being punched or maced). In contrast, when sign language users get the urge to swear in sign language, that counts as coprolalia. For example, one twenty-three-year-old woman learned sign language at seventeen, around the same time that she developed verbal and motor tics. She would find herself signing "fuck" and "shit" while making high-pitched "fu" and "sh" sounds.[7] But that kind of linguistic tic counts as an entirely different symptom from the desire to flip someone the bird or make the "wanker" gesture.

!

Although there's been plenty written about the range of different urges that TS patients experience, it hasn't got us very far in understanding why some people experience these debilitating and hard-to-control compulsions. While on the surface, the urge to shout "motherfucker" in a library might seem a million miles away from the urge to blink or gesture, there is probably a common underlying cause. Whereas TS was once thought to be a psychiatric condition, we now know it's more likely to be, primarily at least, a movement disorder. All of the tics that TS sufferers experience might come from an inability to suppress unwanted, involuntary

movements, even those that are as complex as those needed to say (or sign) a swear word.

That raises a further question: why do TS sufferers find it so difficult to suppress these troublesome urges? Most of us have felt the impulse to do something totally inappropriate at one time or another, like shout "fire" in a crowded theater or touch a flame, but most of us can move past those urges without acting on them. Not so for TS sufferers, or at least not without a great deal of stress and anxiety. Once these urges begin, only performing the action gives relief.

What is so different about the brains of TS sufferers? The answer, unfortunately, is that we still don't precisely know, partly because the syndrome is so variable. As well as the diversity of different types of tic, people with TS process their urges in very different ways. Some find their urges harder to control than others. Some are more likely to act on their urges at times of great stress whereas others find that fear of people's reactions motivates them to control their tics. In fact, some patients try very hard to suppress their swearing in front of their doctors, adding to the difficulty of learning about this condition.

One of the greatest challenges in properly documenting and studying TS is that it rarely strikes alone. Most young people with TS are also diagnosed with depression or anxiety disorders (hardly surprisingly) and are also very likely to have OCD or ADHD.[8] What's more, several of the characteristics of OCD overlap with those of TS. For example, one of the compulsive thoughts that characterize OCD is the urge to shout an obscenity in church, according to the fourth edition of the American Psychiatric Association's *Diagnostic and Statistical Manual of Mental Disorders* (DSM).

We do know that TS is highly heritable: there are genetic variants that predispose children to TS, and those genetic variants are much more likely to occur in children whose family members include other TS sufferers. The exact genes haven't been identified yet because comparing the genetic makeup of TS sufferers takes time, though there are some tantalizing hints as to which gene variants might be involved. To complicate matters, some of the genetic traits that are regularly seen in TS sufferers are also common among sufferers of OCD, which could explain why the two disorders so often go hand in hand.[9]

The effects of TS are hugely debilitating for some sufferers. Motor tics and other compulsions don't just draw unwanted attention; they can often be physically destructive. Muscular spasms commonly cause broken bones, neck injuries, and concussions.[10] Add to that the fact that many TS victims suffer from the compulsion to pick, pull, or otherwise damage their skin—and motor tics are quite obviously harmful. Yet for those patients who experience coprolalia, coprographia, and copropraxis, the physically injurious motor tics aren't anywhere near as distressing as the socially inappropriate urges.

That might sound surprising: who would choose a bit of embarrassment over a concussion or broken bone? But vocal outbursts and startling gestures can cause all sorts of problems ranging from exclusion from school to becoming a target for bullying or abuse.[11] The sad fact is that, for most TS sufferers, the syndrome is at its most severe in late childhood and early adolescence, at the exact time that most of us are making lasting friendships and working out who we want to be. The social stresses caused by TS lead to

problems long after the syndrome itself has become less of a daily burden.

Professor Christine Conelea surveyed 970 adults who had had TS as children and found that many of them still suffered from social exclusion and other psychological difficulties, leading to a poorer quality of life: the lasting impact of TS on children's social and educational lives is distressingly severe. Professor Conelea surveyed 270 children with the syndrome, along with their parents. Forty-three percent of the families said that they avoided social events because of tics, while 38 percent said they avoided public places altogether. That might be because three-quarters of the families surveyed said that they had experienced discrimination and 14 percent had been asked to leave a public place. As if this social stigma weren't bad enough, TS can have a serious effect on education, with 65 percent of children saying that the disorder interfered with their ability to study and 21 percent having been asked to leave a classroom or other educational environment because of their disorder.

Dr. Ruth Wadman of the University of Nottingham has been studying the relationships between children and young people with TS and their peers without the syndrome. Here too, she has found that children with TS are likely to suffer from social isolation and embarrassment, are more likely to be bullied and withdrawn, and are more likely to be rated as less likable by the people who know them.

So much of this distress and isolation comes from a lack of knowledge about the problem. Perhaps we should be concentrating as much on developing greater understanding of the condition as looking for a cure. In the meantime, supportive therapy can help some young people feel more

comfortable in their own skin. Dr. Wadman interviewed six teenagers with TS and found that they all had different coping strategies.[12] Some young people try to make their tics a part of their identity, while accepting that this might make other people feel uncomfortable, essentially saying "fuck it!" to the idea of trying to fit in by suppressing their urges. For some, this is a good solution. One teenager told Dr. Wadman that he deals with his TS "extremely well." Despite his severe symptoms and the depression that he also suffers, he's chosen to let other people decide if they accept him or not. "I basically have decided somewhere in my head that if people can't deal with it, it's not my problem," he says.

But for others it's not easy to be so sanguine about what other people think. For example Hayden, who has moderate TS and OCD, said: "Every time I meet someone new it's almost as if I try and be funny and make them laugh. But that's just me covering my tics, and then everybody just thinks I'm weird from that moment on." With more than one in two hundred children diagnosed with TS, that's a lot of lonely and isolated young people out there.

Because of a general lack of public awareness—exacerbated by the fact that "Tourette's" is so often the punchline to an unfunny joke about swearing—the behavior of patients with TS can seem extremely strange to those of us who aren't familiar with the condition. As well as the unpropositional swearing and involuntary muscle movements, TS sufferers are also likely to experience other compulsions that can get them into serious trouble. Professor Madeline Frank and her colleagues at the University of Birmingham and University College London found that three-quarters of TS patients have impulse-control problems, ranging from uncontrollable

bouts of anger and aggression to compulsive shopping, shop-lifting, hair-pulling, and pyromania. Fewer than one in ten members of the general population and less than a third of all psychiatric in-patients suffer the same problems.

These compulsions are harmful as well as distressing. Nine out of ten of us have no problem controlling the occasional self-destructive urge, usually limited to resisting telling the boss to go fuck themselves. But as difficult as that can sometimes be, most of us have no problem suppressing those impulses when they arise. For TS sufferers, suppressing destructive urges is much more of a challenge and we aren't yet sure why. It might be that the ability to suppress harmful impulses is somehow weaker in TS patients, which would explain why their tics are so hard to control. On the other hand, it might be that the constant effort to control their tics tires out the impulse control parts of the brain.

We all do this: it's a phenomenon called the "ego depletion model of self-control" and it was first studied by Professor Roy Baumeister at the University of Florida. He set up experiments that would force people to exercise impulse control and then tested to see whether they were more or less able to suppress unwanted behavior as a result. In one instance he showed people a film that would make them either happy or sad, but told them to suppress their feelings. Those volunteers gave up much faster on a subsequent test of physical stamina than the other volunteers who watched the same film without trying to suppress their emotions.

He also ran an ingenious test of self-denial and our ability to persevere with difficult tasks in the face of adversity. Professor Baumeister offered volunteers a choice of chocolate or radishes as a snack when they arrived at the

lab. Bizarrely enough—and to me this is the finding that really needs explaining—some of the volunteers were persuaded to choose the healthy radishes over the chocolate. He then gave all of the volunteers a frustrating (and secretly impossible) math problem and measured how long it took for them to give up. The people who had already experienced the frustration of eating radishes (radishes!) instead of chocolate gave up after less than ten minutes while the chocolate eaters carried on for nearly twenty minutes. Could this experiment simply prove that chocolate makes us happier than radishes and thus more likely to persist in adversity? Or that blood sugar levels are correlated with persistence? Possibly, but those volunteers who were offered neither chocolate nor radishes carried on for even longer than twenty minutes!

Even having to *consider* whether to make a self-indulgent choice causes some fatigue to our inhibitions and self-control, both of which are essential ingredients in our willingness to persevere in doing something frustrating but necessary.[13] Which is why I don't even think about dieting.

Because of the complicated but common relationship between TS and impulse control, it's long been hypothesized that the syndrome might be an executive function disorder. Executive function is what we call our ability to switch between tasks, make plans, or use our working memories. Professor Rebecca Elliott of the University of Manchester says that executive function is a bit of a catch-all term for complex processes that coordinate the more basic workings of the brain in order to achieve higher-order goals.[14]

Several things can impair executive function, from fatigue to frontal lobe damage. In order to discern the

links between executive function, impulse control, and TS, Professor Clare Eddy and her colleagues at the Barberry National Centre for Mental Health in Birmingham, England, asked forty patients with TS and twenty non-TS controls to take part in a series of tests of executive function. Crucially, all the volunteers with TS were otherwise healthy: they didn't have ADHD, OCD, or other psychological or neurological problems, so she could study the effects of TS in isolation.

Executive function includes things like verbal fluency, which Professor Eddy tested by asking the volunteers to name in one minute as many words beginning with the letters F, A, or S as possible: a particularly provoking set of initial letters for English-speaking coprolalia sufferers! She also tested their working memory by having them listen to three- to eight-digit numbers and then recite them back in ascending order. So for example, if you heard 76843 you would have to reply with 34678. Finally, she tested impulse control with something called the Stroop test.

The Stroop test is a simple psychological experiment. Volunteers are shown the names of colors, printed in colored ink. For example, I might show you a card with the color "red" named on it, but printed in the color blue, or "black" printed in yellow, "green" in orange and so on. Your job would be to name the color of the ink rather than saying the name of the color written on the card. It's surprisingly difficult, and we can quantify exactly how difficult people find this task by comparing their reaction times when the name and the ink match ("red" in red ink for example, known as a "congruent condition") and when

they don't match ("red" in green ink, an "incongruent condition"). If the colors match, the average person can correctly name a hundred colors in sixty seconds. In contrast, for the incongruent examples it takes on average 110 seconds to name a hundred colors.[15]

The Stroop test allows us to probe exactly how difficult it is to inhibit an easy-but-wrong response in order to allow a difficult-but-right one to come out. The bigger the difference between the response times, the harder the brain is having to work in order to exercise its inhibition.

Professor Eddy and her team found that TS patients could generate on average forty entries on the "list words beginning with F, A, or S" task, whereas the control group generated fifty. She also found a slight difference between the number-ordering task: TS patients could manage entries up to six digits in length, on average, whereas the control group could manage almost seven digits. On the Stroop test, TS patients made one and a half as many errors (and took longer) than the control group.

Patients with uncomplicated TS found all three tests of executive function much harder than the healthy controls, but this could mean one of three things: that TS causes the executive function problems, problems with executive function cause TS symptoms, or executive function impairment and TS are two separate disorders that just happen to show up together frequently like TS and OCD. According to Professor Eddy, we still don't know which explanation holds; TS sufferers definitely struggle more with inhibition, but no one knows why. However, it is possible for many TS sufferers to delay or divert their reaction to those urges.

!

While some patients find it easier to suppress their swearing in public (where anxiety levels are high, but the social penalties for swearing are higher), others find it easier to suppress their swearing among friends and family, in settings where they are more relaxed. Stress might increase the severity of the urges that some TS patients feel, which would explain the greater severity of tics in high-stress situations. For others, the fear of social consequences acts as a very strong motivator to suppress swearing and other tics, no matter what the cost. To make sense of this, Professor Conelea set up an experiment to differentiate between situational stress and social pressure. She set TS patients time-limited mental arithmetic problems in order to induce a baseline level of stress. She told these patients that they could swear or tic as much as they wanted. When allowed to tic freely, these patients' tic rates were slightly lower while doing the maths challenge than when they were just resting, possibly because of the amount of concentration that the mental arithmetic required. She then asked them to try again, but this time to try to suppress their tics. As a result, the number of tics per minute almost doubled under stress. The very stress of trying not to act on ticcish urges made it much more likely that the volunteers would suffer from tics.

Behavioral studies like these can tell us a lot about the environmental and psychological factors that affect patients with TS, but they still don't tell us what's going on inside the brain. The behavior of each and every one of us is the product of a complicated interplay between chemical and electrical signals, but we still have only a vague

understanding of how these all fit together. New brain-imaging techniques and a better understanding of the complexities of TS sufferers' behavior mean that the picture is now clearer than ever, but we still lack a complete model of what makes TS sufferers tic.

One possible reason why Tourette's syndrome sufferers develop tics has to do with an important chemical in the brain. We believe dopamine plays an essential role in TS, mainly because the drugs that most successfully treat the disorder have an effect on the way the brain uses dopamine as a signal. Dopamine is a neurotransmitter—a chemical that neurons release in order to encourage other neurons to fire—that has a range of functions in the human brain and the body. We need dopamine to tell the kidneys to excrete urine, and to tell the pancreas not to overdo it on the insulin production.

In the brain, dopamine is yet more of a shape-shifter. Even though, of the 100 billion neurons in the brain, only 20,000 or so are receptive to dopamine, these 20,000 receptors have a huge influence on our behavior. We experience the action of dopamine in certain parts of the brain as a reward. When you do something difficult that pays off, the high you feel is, in part, the feeling of dopamine being released by some neurons and absorbed by others. Too much dopamine can make you psychotic, too little and you lose the motivation to try difficult things because you don't receive the dopamine "high" you expect as a reward. Cocaine gives you an "unearned" dopamine hit, while antipsychotics stop dopamine from giving us that sense of reward at all.

The action of antipsychotics on TS gave the first indication that the brain's reward system might somehow be

involved in the syndrome. It wasn't until 2011 that Professor Stefano Palminteri and his colleagues at the Pierre and Marie Curie University in Paris studied the effect with a clever test that examined the role of rewards in people with and without TS.[16] His recruited volunteers were to carry out a test of how rapidly they could press a combination of keys when they saw a prompt on a screen. The volunteers were asked to put the fingers of one hand on the C, F, T, H, and N keys of their keyboard. This puts all five digits in a slightly awkward inverted V-shape. The screen would then flash up a prompt, telling the participants the order in which they should press the keys.

Each volunteer saw ten different sequences and each sequence was shown fifteen times over the course of the experiment. Of those ten, five would give them a maximum ten-euro reward if they pressed the keys correctly, while the other gave a maximum reward of one cent.

Imagine that you were told that you'd get ten euros if you typed the sequence CFTNH quickly and correctly, but only one cent for FTHCN. Consciously or not, you'd expect that your performance on CFTNH would be better than FTHCN by the end of the experiment. And you'd be right: for most of us, dopamine is the neurotransmitter that kicks in to strengthen memory traces when the reward for an action is high. The performance of those volunteers who didn't suffer from TS improved more on the ten-euro sequences than the one-cent sequences, with about a quarter-second speed advantage when it came to the high-reward key presses.

Among the volunteers with TS, the results fell into two distinct categories. For those TS patients who were being

treated with antipsychotics, there was very little difference between the five "expensive" sequences and the five "cheap" ones. They improved at both with practice (though could still barely manage to shave off one-twentieth of a second from either), but they didn't make any extra effort, consciously or unconsciously, to get the high-reward sequences right. However, for the TS patients who weren't taking medication, the difference between the high- and low-reward conditions was massively exaggerated. While the presence of a higher reward inspired the volunteers without TS to make a quarter-second reduction, the unmedicated TS patients shaved a whole second off their performance times in the high-reward condition.

Exactly how they did so is not entirely clear, but it's reasonable to suppose that the higher levels of dopamine in the unmedicated TS sufferers' brains meant that they experienced a much more significant neurological reward than the other two groups. This could explain why tics proliferate and become stronger over time: performing the tic satisfies the urge and causes a release of dopamine. The exaggerated rewarding effect of dopamine would make TS patients overlearn the movements that make up the tic. It might be the case that performing a tic, acting on an urge, triggers the release of dopamine in a TS sufferer's brain—to such a degree that suppressing that urge might be exactly as painful, and for the same reasons, as refusing a line of coke would be to an addict.

For those of us without TS, the neurological rewards we experience from the release of dopamine are strong enough to motivate us to do difficult things that are likely to be of some benefit, but the rewards aren't so strong that we can't

suppress our urges where necessary. For those people with TS, that exaggerated dopamine hit that makes it possible to crush the opposition on a simple key-pressing task might be the driving force behind the ticcish behaviors that are so hard to subdue. Before we know for sure, we need to understand much more about the brain, but in the meantime such research does offer a better picture of how TS sufferers' brains are special, as well as how antipsychotics might work.

!

We've known since the 1950s that treatment with antipsychotic drugs can alleviate the symptoms of TS, so why aren't all TS sufferers simply offered antipsychotic medication? After all, it was the effectiveness of treatment with antipsychotic drugs in the 1950s and '60s that first made it clear that TS is biological rather than psychological.[17]

The mistaken belief that TS was psychological led to some interesting and misguided "cures" over the years. In 1957 the British Medical Journal reported on the (in hindsight) extraordinary treatment given by Dr. Richard Michael of the Maudsley Hospital in London to a man in his late twenties after using cutting-edge psychoanalytical approaches. Dr. Michael looked at the man's home life (strong mother, weak but caring father) and sexuality (the "active participant" in same-sex relationships during his army service) and decided that the patient's need to repress his sexuality in civilian life was probably the issue.

The treatment administered by Dr. Michael was "carbon dioxide therapy," in which the unfortunate patient was made to breathe air that was 70 percent CO_2. Just to put that in perspective, that's nearly twenty times the concentration

of carbon dioxide we exhale during normal breathing. After thirty sessions of being gassed—during which Dr. Michael says he observed "violent sucking movements" and dreams that were "full of obviously phallic imagery"—the patient reported that he was pretty much fully cured, thank you kindly! Whether this was due to the efficacy of the treatment or simply the desire not to be gassed is open to question.[18]

What we now know, after more than a half-century of pharmacological treatment and hundreds of studies, makes this approach seem baffling, bordering on the inhumane. That said, antipsychotics, while helpful, can sometimes be equally unpalatable.

Antipsychotic drugs do help to reduce tic severity for some patients, but they aren't suitable for all. Many patients are overwhelmed by the side effects, which can include headaches, dizziness, sedation, depression, weight gain, and, in some cases, symptoms similar to Parkinson's disease. Even for those able to tolerate them, these drugs aren't a permanent solution: discontinue taking them and the tics resurface.

It's no wonder that many patients are still searching the fringes of medicine for treatments that, while much less likely to provide benefit, don't lead to tremors, pain, and misery. According to Professor Diana Van Lancker of New York University, Botox shows some promise as a way of reducing vocal and muscular tics.[19] It seems to reduce the strength of the premonitory urges that many TS sufferers experience and each injection is effective for around three to six months. Professor Van Lancker suggests that the Botox reduces the amount of muscle tension in the vocal cords, thus reducing the compulsion to form a swear word, but there's still not enough research to show whether this is a viable treatment, or just so much more CO_2 therapy.

A number of dentists have also started offering an implant that works on a similar theory: that muscular tension in the jaw is responsible for the tics in TS. There haven't been any clinical trials yet but Dr. Andrew Clempson and his colleagues from the charity Tourette's Action contacted their members to see whether they could find people who had experienced any relief after a dental implant. They found only nine people who had tried it, but each of these had paid between £3,600 and £10,000 for the process. Three of the patients experienced some relief, two had no relief and the other four said they were experiencing ongoing complications. According to Dr. Clempson, patients should save their money, at least for now. "There is no sound theoretical basis for dental [treatment] for TS. Our small survey suggests the treatment is less successful than sometimes claimed," he said. But watch this space, as the Tourette Syndrome Association is planning a thorough clinical trial soon.[20]

If a £10,000 implant in your jaw doesn't sound appealing, what about a £30,000 implant in your brain? Deep brain stimulation (DBS) is a relatively new technique in which electrodes are implanted in structures deep inside the brain. Since the early 2000s, DBS has been used to treat Parkinson's disease, OCD, depression, and chronic pain, and in more recent years has been tried in a number of patients with TS.

Because the surgery is so expensive and so complicated, the number of patients treated to date is small, but most trials show that there is some beneficial effect to stimulating the thalamus, a structure found deep in the brain of almost all animals, that coordinates sensing, movement,

consciousness, and sleep. While intermittent stimulation given to both sides of the thalamus seems to lead to a reduction of tic severity of around 70 percent, the procedure is not without its drawbacks. Patients are at risk of stroke, infection, strange sensations elsewhere in the body, lethargy, and problems with their vision.[21]

One Dutch study was halted halfway through recruitment because all of the patients taking part reported debilitating side effects.[22] In another study, one patient died after her symptoms worsened, leaving her unable to swallow. At the age of twenty she had DBS electrodes implanted, but as soon as stimulation was switched on her tics increased and she became gripped with serious anxiety and the overwhelming urge to stab and scratch herself. She asked for the stimulation to be switched off but, even after the current stopped flowing, she still suffered complications and could no longer bear to swallow food or drink. She became so depressed and withdrawn that, at the age of twenty-three, she died in a nursing home from severe dehydration after refusing to be treated with intravenous fluids.[23]

These experimental approaches raise some serious questions: most newly diagnosed TS patients are children and adolescents, and many suffer from the compulsion to self-harm. Is it ethical to submit these patients to an experimental procedure that can cause such serious complications? According to Dr. Roger Kurlan, medical director of the Movement Disorders Program at Atlantic Neuroscience Institute in New Jersey, surgeons should take the utmost care when choosing whether to treat a patient, no matter how severe their tics, since it's impossible to predict the severity of the complications they might suffer as a result.[24]

!

So far, then, no treatment option seems particularly desirable. Drugs have debilitating side effects, Botox and dental implants are poorly tested, and deep brain stimulation is still in its shaky infancy. Dr. Sabine Wilhelm of Massachusetts General Hospital and Harvard Medical School thinks that it's time to start looking again at behavioral treatments for TS. Because of the mistakes of the early twentieth century, when questionable psychological treatments failed so spectacularly at helping TS sufferers, it's no surprise that it's taken fifty years to revisit the idea of therapy as a serious option. But in the early 2000s, as we started to understand the extent to which our behavior can shape our brains as well as the other way round, several researchers have chosen to reexamine the idea of behavioral interventions as an alternative to medication.

Until they took on this project, it was common wisdom among doctors that the tics in TS simply couldn't be controlled, that trying to suppress a tic would just cause new ones to rise up and take their place and that behavioral therapies would only worsen the symptoms. But a new approach known as comprehensive behavioral intervention for tics (CBIT) is proving to be surprisingly effective.

According to Dr. Wilhelm, the success of CBIT means that it's time to ditch resistance to behavioral approaches. Researchers and medical specialists alike have been concerned that tics might get worse if such approaches are relied upon, that patients might be unduly burdened by the effort required to undergo such treatment, or that it

represents a return to the "dark ages" of regarding TS as a psychological rather than neurological disorder.

None of these concerns is valid, says Dr. Wilhelm.[25] Many patients are eager to try a ten-week program of therapy that might permanently improve their quality of life, especially when antipsychotics are the only well-tested alternative. Furthermore, our behavior can shape our neurology, given enough reinforcement and support; behavioral approaches could very well change the way that TS sufferers experience their urges and their tics on a physiological as well as a psychological level. Whereas the therapy for teenagers that Dr. Wadman studied was aimed at helping patients to accept and live with their tics, Dr. Wilhelm strongly believes that there are therapies that can change the urges and the tics themselves.

TS is undoubtedly a biological phenomenon, caused by our genetic makeup and the structures of our brains. But, according to Dr. Alan Peterson of the University of Texas, our current understanding of behavioral therapies combines the effects of genes, brain structures, and biology with the environment and the situations that TS sufferers find themselves in. Behavioral therapies like CBIT can help TS sufferers understand their environmental triggers and, in response to their urges, to build repertoires of behavior that they find less stressful.[26] "Behavior therapy for tics is not a cure, but a management strategy that can help people live a better life," he says.

In CBIT, patients are helped to identify the types of situation that make tics and urges more severe, and to be able to notice the premonitory urge early on. The patient is encouraged to rate the repertoire of their tics from the most distressing to the least distressing and treat them in that

order. For each tic, they try to learn another behavior that will both be incompatible with carrying out the tic, and won't be socially or physically damaging. For example, they might try deep and steady breathing to counter the urge to swear. This habit-reversal training can break the link between the urge and the tic over a surprisingly short period: a few weeks or even days.

CBIT also includes relaxation training, as well as training for parents, teachers, and other important people in the patient's life, in order that they can support the behavioral changes. Parents are encouraged to notice and praise their children when they successfully divert an urge to tic, and to help them avoid situations known to make their urges worse.

The CBIT approach is promisingly effective. In a study of 126 nine- to seventeen-year-olds, the patients who took part in eight sessions of CBIT over ten weeks found that tic severity as measured by the Yale Global Tic Severity Scale reduced by almost a third, a result twice as good as that achieved by patients in the control group who simply met with someone to discuss TS for the same period of time.[27] What's more, two-thirds of the children who responded well to CBIT were still feeling the benefit six months later, and adults also showed similar improvements in a study of 122 people taking part in the same sort of therapy.[28]

Tic reduction of around 30 percent isn't as impressive as the effect of antipsychotic drugs, which can reduce them by 60 to 80 percent in most individuals. But drug side effects leave many TS sufferers trapped between the crushing emotional and social burdens of TS and the debilitating nature of

the medication. Therapies like CBIT offer a middle ground: the hope of at least some relief without that risk.

So far all the research has focused on ways of reducing the frequency of tics, but there's another way to ameliorate the long-term social and emotional impact of TS. The current lack of public awareness of the condition leads to bullying, isolation, depression, and anxiety, so there's one very simple thing that would make sufferers' lives much more bearable: for the rest of us to learn to be a bit more understanding.

For TS sufferers with coprolalia, their compulsions to swear are not like the occasional verbal pyrotechnics that the rest of us sometimes come out with. They are constantly having to control urges that, if acted upon, would release a dopamine hit that most of us have never experienced. Despite this dopamine rush, and for reasons we still don't understand, TS sufferers' tics sometimes seem to be as inappropriate and harmful to themselves as possible. While I get a good dose of painkilling relief from non-propositional swearing, or can make a joke or a point with propositional swearing, the unpropositional swearing in TS makes life uncommonly difficult for the sufferer.

Medication and behavioral therapies can mitigate those urges, but TS patients either have to accept serious side effects, or work very hard at developing partial control of their tics. I just can't think of any other disease where the rest of us expect the victims to increase their discomfort just to make everyone else feel better.

That's what we are asking these TS sufferers to do—and remember TS is much more common in children and adolescents than it is in adults. Essentially, we are offering these children and their parents a lousy choice: control your tics

with difficult and sometimes dangerous methods, or risk bullying, abuse, and exclusion.

So, as you can see, if you're a Tourette's sufferer, swearing really isn't all that good for you, even though it feels great, and that's why this chapter shouldn't be in this book. Tourette's isn't the punchline to a bad joke about swearing. It's a miserable and poorly understood disorder with serious consequences. Dr. Heather Smith of the University of Manchester interviewed several young people about their experiences with TS.[29] One of the teenagers that she spoke to said "I wasn't a person anymore, I was basically just a machine that hit things and shouted." That's a frightening and lonely thing for any child to believe about herself.

What if we all learned more about the condition? What if we decided to accept the fact that these tics are more distressing to the TS sufferer than they are to the rest of us? While that wouldn't be enough to make those overwhelming urges go away, it might at least reduce the number of young TS sufferers who go on to live a life of social withdrawal and depression. I think it's time we tried.

₄

Disciplinary Offense:
Swearing in the Workplace

One of the most widely reported revelations about the 2008 banking crisis had nothing to do with malpractice or negligence. Instead, breathless column inches and earnest talking heads discussed the discovery that a Goldman Sachs senior manager sent an email that described a subprime mortgage arrangement as "one shitty deal." When the news broke, rather than apologizing to the public for the years that they sold shitty products, Goldman Sachs instead announced that they had instigated a rigorous email filter and a "no swearing" policy: doesn't that make you feel better?

Swearing in the workplace divides opinion. TV chef Gordon Ramsay has made it part of his public persona, while Paul Dacre, editor of the *Daily Mail*, uses one particular swear word so much that his briefings are known among staff as the "Vagina Monologues." In other organizations, in the United States especially, a new puritanism has taken hold. More afraid of offending their customers with bad language than bad products, many companies are cutting out

the cursing. But research from around the world, and particularly Australia and New Zealand, shows that—in some cases at least—the team that swears together stays together.

The Academic Study of Banter

Swearing and insults—even ones that can sound quite vicious to the uninitiated—are all part of the banter in many workplaces. It's good for group bonding, and inclusivity makes for a productive workforce. As Dr. Barbara Plester wrote in her 2007 paper, "Taking the Piss: Functions of Banter in the IT Industry": "Banter occurs when people are in good humor; when people are playful, they are at their most creative."[1]

Like me, Barbara used to work in IT before leaving to become a university lecturer. "My husband's still in that industry so that keeps my connection pretty live. [Also] I've got two brothers in the industry so we're kind of an IT geek family." So the types of jocular abuse (piss-taking in the UK) she encountered came as no surprise—to her or to me. What is surprising is the degree of *nuance* that goes into telling someone to fuck off.

Barbara interviewed and observed a mix of men and women at three small IT companies in New Zealand and almost all of them said that they'd taken the piss (and had the piss taken out of them) at work. In fact, piss-taking is a matter of pride and—male or female—most employees were insistent that they were "included" in banter and that they gave as good as they got.

In the three months of observation that Barbara carried

out she discovered many things. Firstly, "piss-taking"—whether it involves swearing or not—is really important for team bonding and morale. The most common piss-takes were all ways of bringing the team closer. The employees of these three companies only insulted the people they knew and got along well with. Being initiated into the banter was a sign of finally being accepted. Whenever someone new arrived, insults would start off gently then ramp up as the team tested to see whether the new person could take a joke.

Barbara does note one startling (for the newcomer at least) exception to the rule. The self-appointed workplace joker "terrorized" a new male employee by inviting him to a game of "all-male nudie leapfrog" being held at lunchtime.

"God I remember that," she says amid gales of laughter, "All-boy nudie leapfrog!" Was the new guy really that terrified? Apparently, he left the company not long after. "That's very Freudian, actually. That's what Freud said—we put things in a joke to protect ourselves and then we can say the unsayable—I love that stuff."

Piss-taking, swearing, and jocular abuse allow people to talk about things like race, sex, and other types of difference in a way that is—or at least can be—a sign of solidarity. Alpha Tech has about fifteen employees in New Zealand, but it's part of a much larger international IT group. Barbara was sitting in on a Monday morning teleconference.

"It was just the general banter going on—I was one of the gang by then—we were sitting round the tables and, you know how the technology always takes a while to fire up . . ."

One of the employees, Kara, decided she'd had enough of the men always running the videoconferencing equipment and snatched up the remote control. Alf, one of her

colleagues, started calling Kara the "evil, remote-control woman." Far from being offended, Kara reacted with glee, taking it as a sign of recognition that she'd wrested control over the equipment from the men. As a woman in a male-dominated profession, that kind of joshing is a sign of acceptance, of finally becoming one of the guys.

Did that happen to Barbara? And, as a scientist, would she give as good as she'd got? Not quite, she says, but it's a fine balance she had to achieve.

She described how hard this could be:

> If you sit back and are removed from it, playing the objective researcher role, you kill the humor. You've got to just loosen up and gently become part of it without being the initiator . . . You end up having to use a bit of the language in the right situations. If someone played a prank on you the standard reply was "fuck off"—you had to respond with that or you were just a patsy, really!

Even when it doesn't explicitly involve swearing, jocular abuse (or, less technically, "piss-taking") hits a lot of the same spots as bad language. Usually, piss-taking involves a taboo, or at least some topic that the piss-takee might be expected to find sensitive. That's why height jokes tend to be aimed at the short (or the *extremely* tall). Piss-taking is also designed to get an emotional response out of the hearer. While I can't find any studies on the effect of piss-taking on heart rate and galvanic skin response, it's my intuition that, along with the laughter, the piss-takee (and probably the piss-taker and bystanders) experience quite

a bit of emotional arousal. As with jocular swearing, piss-taking requires a lot of trust and bonding in order to be sure that it's going to go over well.

What the researchers in Barbara's team found is that it all comes down to relationships:

> [Piss-taking] is such a relational device between people that if the relationship is OK you know what you can get away with, which boundaries you can cross and which ones you can't. It's not really predicated on race or gender or that sort of thing—it all comes down to how well people know each other.

A lot of the jokes involved swearing or at the very least they challenged taboos. "Many of the jokes are outright racist, sexist, or otherwise personal." Fale, a Samoan New Zealander, talks about her European New Zealander colleague whom she calls an "FOB" (fresh off the boat). "I make English jokes about him and call him 'fat boy.' Everyone laughs." Doesn't it bother Fale? Apparently not. In an interview with Barbara she laughed and said that it was "fun" and that she "gave as good as she got."

Of all the insults and jokes that Barbara observed, very few seemed to overstep people's boundaries. Some people were never teased about their weight, for example. Barbara noticed that "fat jokes" were leveled only at people who made the same joke about either themselves or others. The backslapping, laughter, and smiles that went with these jokes meant that they didn't *appear* to cause offense. In order to come off well, jocular abuse either has to stay within the boundaries that people set for themselves by making their

own self-deprecating jokes first, or it has to be so outrageous that it can't possibly be meant seriously.

Barbara had expected that jibes about race, sex, and all the other modern taboos that come under the umbrella of political correctness would be treated with extreme caution but instead the insults were raucous, risqué, and reciprocal. There's an odd effect at play with some of these racial insults—at least in theory. Research conducted in the 1970s suggests that the more outrageous the insult the more intuitively it is construed as a joke, whereas milder insults are more likely to be heard as "meant." The incongruity, that "gasp" moment between the leveling of the insult and the joke being accepted as such, is very much like our responses to swearing. The response is the emotional followed by the intellectual. The picture is even more complicated because so many joking insults include swearing and the bigger and bolder the broken taboo, the higher the stakes but the bigger the joke.

The Banter Backlash

That's not a message that's welcomed by writer and motivational speaker James V. O'Connor. In 1998 he began a one-man crusade to cure American workplaces of the vice of swearing when he founded the Cuss Control Academy of Northbrook, Illinois.[2] For a payment of $1,500 an hour to his organization, Mr. O'Connor will bring his brand of "humor and overlooked common sense" to your organization. He operates from a leafy suburb on the edge of Lake Michigan. His home office lies in a triangle bounded by a nature preserve and not one but two adjoining golf courses—there's

a certain defiant gentility to both the place and his manner. In his measured and slightly folksy way, Mr. O'Connor talks proudly of the corporations, societies, and schools that he's visited in order to purge profanity.

"Some companies hired me because there was a real problem with swearing in the company and they didn't want to single anybody out, so they just had me come in and do a presentation to the entire staff." American audiences really enjoyed being taught not to swear, he says.* I try not to sound too skeptical when I ask if that even includes the high-school kids.

"Oh yeah!" he says, proudly. "I spoke at a lot of high schools. I made it funny. In one school I got a standing ovation—2,000 students gave me a standing ovation."

Getting 2,000 hormonal teenagers to appreciate any-thing, let alone being instructed not to swear, seems little short of miraculous. How does he manage to get kids that excited about clean language?

> What I do is I say to kids: whenever you're mad you're either . . . [at this point he pauses, and seems to handle the next few words with rubber gloves] You're either "pissed off" or "fucking pissed off"; there's nothing in between. Think of other words that you could use. They came up with "mad," "outraged," "upset," "livid," "furious," "perturbed" . . . I said, if you say to your friend "I'm really perturbed" he'll probably laugh because it's kind of a funny word. But if you say "I'm really upset" he'll probably be more interested in hearing what your problem is than if you say you're "effing pee owed."

---

* It takes all sorts, I suppose.

It sounds like a fair point—I personally think "perturbed" is a fantastic word—but I'm still not quite getting the standing ovation. I ask him why there's such an appetite for self-improvement via swearing elimination in the States. A hangover from their Puritan forebears, perhaps? According to Mr. O'Connor the cause is much more recent than that. He believes that swearing is a modern ill.

"It's part of our casual society: casual clothes, casual sex, casual relationships with people who used to be called Mr. and. Mrs. In the past you never called the boss by his first name." Mr. O'Connor (not James; definitely not James) warms to his down-home, common-sense oration. "The 1950s and '60s were very formal and rigid, but along came the civil rights movement, the antiwar movement, women's liberation; a whole lot of things going on to make people feel more free to be who they wanted to be."

Mr. O'Connor thinks that things also changed when women entered the workplace. Decades ago, he says, if men swore at all they certainly didn't swear in front of women. But then women made it into the workplace, "and they had to be one of the guys; they had to act like men, dress like men, wear suits and everything else, and try to talk tough, and they thought that had to include swearing."

Is this what the fight against swearing comes down to? A nostalgia for a simpler, happier time? There are so many questions that I want to ask him about the points he's just made, but his courteous certainty makes me lose my nerve. I thank him, wish him all the best, and scurry back to the twenty-first century, where the language might be casual but at least I get to wear trousers if I want to.

Some might yearn for a golden age of clean language

(which, as we'll see in the next chapter, probably never existed), but what about life in real twenty-first-century workplaces, where women and men from different cultural backgrounds work together? Well, thanks to Professor Janet Holmes and her colleagues from the Language in the Workplace Project at the Victoria University of Wellington, we don't have to guess.

An Appreciative Inquiry into Swearing

Professor Holmes is a well-spoken New Zealander with a passion for understanding how people communicate. Rather than deciding how people in workplaces *should* talk, she set out to study how they *do* talk. Her team goes into factories and offices to see how humor, small talk, and swearing play a role in the work environment. "Our team's approach is what's called an 'appreciative inquiry.' We look for what people are doing well. We were called in by [one company's] HR people because they had this particularly high performing team and they wanted to know what made it work."

This team of sixteen men and four women—nicknamed the "Power Rangers"—is a tight-knit group of Samoan, Maori, Tongan, and European New Zealanders. They agreed to carry voice recorders around for a week. The swearing just jumped out of the transcripts, says Janet.

To an outsider, all this swearing might seem off-putting —as though the Power Rangers are a really hostile bunch of people. But again, Janet stresses that there is a nuance to swearing that is extremely important. Thankfully, she is willing to demonstrate the different uses of the word

"fucking" in the way that only a professor of linguistics can:

> Generally speaking swearing has a number of different
> functions. One is to just emphasize what you're saying;
> it can act like any intensifier, like "really" or "very." But
> "fucking" is used so regularly in this particular context
> that it's part of the jargon of the group. You can tell
> from the way they pronounce it whether it's likely to be
> aggressive or whether it's just an intensifier. Often it's
> to do with the volume and the stress like, "not fucking
> likely" as opposed to "not FUCKING likely."

I can really hear the difference. Possibly her colleagues
in the adjacent office can, too. "Y'know," she reflects, as I
hold my breath and wait for someone to knock on her door
to check whether everything is OK, "the louder it gets, the
more of a signal it is that the person is actually annoyed as
opposed to using it just to take it up a notch."

The paper that she and her colleagues published in 2004
contains no shortage of swear words.[3] Mr. O'Connor would
be mortified. But all that swearing kept the team bonded,
particularly when it came to whinges or complaints.

If whinges and complaints sound like the same thing
to you, then you're not a professor of linguistics. Janet
explains the difference: a whinge is the kind of grumbling
that you indulge in when you don't expect anything to
change ("fuck this shitty weather") whereas a complaint
is a grumble with an implicit request for change ("I wish
people would stop trailing their wet umbrellas all over the
fucking floor"). Complaints, she explains, are a "face-
threatening act."

That sounds pretty violent, but actually a "face threat" is

entirely metaphorical. Anything that might cause someone to "lose face" is technically a face-threatening act, so complaining about work or questioning someone's competence is a face threat.

Take this exchange between Russell,[*] a Samoan-European in his late twenties and Lesia, a Samoan in his early thirties:

"[I'm] fucking sick of this line," says Russell to Lesia. "Stuck here all the time."

"If I put you on that line you're getting worse," Lesia replies. "Fucking worse . . . Slow like an old man."

To me that sounds like fighting talk. But again my instincts are wrong. Janet's data show that the insults and swearing are the oil that greases the Power Rangers' wheels. "It's very friendly. It's not a bitter complaint. [We found] that swearing was clearly one of the ways in which they expressed solidarity with each other and that they got on well."

Swearing for solidarity is something that Ginette, the Power Rangers' team leader, does very well. Although she's responsible for making sure the Power Rangers work together as a tight unit, she's very much "first among equals." Much of her time is spent dealing with whinges and complaints like these. But Ginette has a secret weapon: she's an expert in understanding how and when to use swearing.

Janet becomes animated, obviously remembering Ginette with fondness:

Ginette speaks fluent Samoan. One of the interesting things about her was that, at the team meeting at 6 a.m. every day, Ginette was very direct and used really "in yer

---

[*] All names here are pseudonyms.

face" language. She would lambast them if they hadn't achieved yesterday's objectives; tell them they had to do better today. But afterwards, when she went around and talked to the team members one by one, she was much more motherly. She even used Samoan to the Samoan speakers to check that they'd actually understood everything and were on board.

On the rare occasions when the Power Rangers did fuck up, Ginette would tell them to their faces. But she would also defend them elsewhere in the company using language that was much more polite and diplomatic. Ginette sounds very fluent—not just in English and Samoan but in swearing and not swearing as well.

In fact, Ginette sounds like the kind of woman in the workplace who would give Mr. O'Connor a fit of the vapors. According to Janet, however, she is an extremely astute communicator—and the swearing is just part of that. "As most people do—depending on who they're talking to and in what context—she changed her style. So she was very abrupt and direct, and she could abuse people if they shouted comments to her in the morning meeting, but then afterwards, one-on-one, she was much more jokey and sympathetic."

Swearing in the Power Rangers team is obviously full of nuance. What would happen if a newcomer tried this? If I joined the Power Rangers tomorrow, would I have to start swearing straightaway? To my delight, Janet has looked at this problem, too, studying the experience of immigrants who have recently arrived in New Zealand. The important thing is to prepare newcomers so they don't get a shock, she says:

It's one of the things we thought about quite a lot because we're working with migrants who were coming to work in factories and construction sites—that's the other place where we found a lot of swearing, on building sites [to the surprise of no one at all], and so we take the view that you should warn them that they're going to hear this language.

Many new arrivals find it shocking to hear this kind of language in their workplace. It's not that they'd never swear, but many migrants come from cultures where swearing almost always has negative connotations. In that case it can be hard to get used to your colleagues doing it. And, as we've seen, even if swearing isn't necessarily thought of as vulgar by default, what counts as an unforgivable insult in one culture might be laughable in another. Getting the level of offense right is a tricky business:

> The first thing is just to say: don't react negatively to this. Um, and . . . [Janet pauses, choosing her words with care] there's no *requirement* that they use it—but if they want to be integrated into the team it's gonna take them longer if they resist using the language themselves. I'd say: if you can use this sort of language yourself comfortably then fine, but otherwise just wait until you feel you can do it. You're not likely to be able to change the workplace if you're a lone voice among a whole group who have developed their own ways of talking.

Brits and Kiwis have much in common when it comes to swearing at work, but the transition, even for a native-born

Brit, can be a difficult one to navigate. Take the case of British researcher Stuart Jenkins who found exactly that when he went to work in a mail-order packing warehouse in East Anglia to help pay for his studies. The line Stuart worked on didn't have a Ginette to help gel the team together. Instead, the informal leader on the floor was "Ernest"—a large and boisterous chap who enjoyed making temps in general and students in particular as miserable as possible, tripping and "play-punching" them in the stomach. After a couple of months of this treatment, Stuart had had enough.

One afternoon on the packing line, Stuart was suffering under a barrage of insults from Ernest. Ernest kept insisting that Stuart was a lightweight and that he, Ernest, always managed to take on far heavier duties.

"Well, fucking get on with it then, you lazy cunt," Stuart shot back.

The workers around them made the kind of noise usually associated with mechanics and plumbers—the noise that is the universal signal for "this is going to cost you." Ernest muttered an expletive-laden insult but walked away. Anyone unfamiliar with Janet Holmes's research might have advised Stuart to watch his step after that, but instead he found himself being invited to the pub that very evening and from that point on he was part of Ernest's "in group."

Swearing really can break down barriers. But of course, even among workmates swearing and abuse aren't always taken well. What really matters is whether or not the joke has a "side" to it. Barbara Plester describes this type of swearing as a "barbed message." For example, when one of the participants in her study, Kara, didn't buy Alf tickets for a company cinema trip he was upset at being left out.

He burst into the room with the words, "You're a bitch," and later when he left repeated, "Karl came through for me [with tickets], but you're still a bitch."

The thing that sets Alf's outburst apart from the other banter—the reason why Kara might have felt offended by it even though she reveled in being called the evil remote-control woman by Alf—is because Alf genuinely was upset with her and wanted to let her know. By disguising his outburst as just part of the usual office joking he not only vents his anger, but also is able to wave off any comeback by claiming that he's "only joking." Barbara says that these outbursts are a way of the complainer trying to eat their cake and have it too, linguistically speaking.

We do that here in the UK, too. Dr. Michael Haugh from Griffith University in Australia and Dr. Derek Bousfield from the University of Central Lancashire compared what they called "jocular mockery" between British and Australian groups of men.[4] Brits in particular use pointed swearing to puncture what looks like an inflated ego.

The Importance of Taking the Piss

As Kate Fox observes, in her funny and incisive work of social observation *Watching the English*:

> In most other cultures, taking oneself too seriously may be a fault, but it is not a sin—a bit of self-important pomposity or overzealous earnestness is tolerated, perhaps even expected, in discussion of important work or business matters. In the English workplace,

however, the hand-on-heart gusher and the pompous pontificator are mercilessly ridiculed—if not to their faces, then certainly behind their backs.[5]

This type of ridicule usually consists of "taking the piss"—a particular kind of barbed mockery that is meant to stop people getting too cocky. In the UK, boasting about your achievements is in extremely poor taste. The only socially acceptable way to draw attention to your triumphs is with a bit of self-directed piss-taking. If you don't do that, rest assured that your mates will. In a transcript that Haugh and Bousfield call "Media Whore," they document the taking down a peg or two of a colleague who managed to get himself on television. This colleague, Simon, took part in the Great North Run dressed as one of the Three Musketeers. His two colleagues—the other two musketeers—were also interviewed with him, but only Simon got a repeat broadcast. He makes the mistake of bragging about this.

"Fucking four times on the fucking telly," he laughs. His mates laugh too and his colleague David replies with an incredulous but appreciative "Fucking 'ell." But then Simon goes on and commits the faux pas* of making himself out to be a bit special:

"An' they only played my bit. They didn't play the lads', did they?" Big mistake.

"You big-headed bastard," breaks in David, making the lads laugh. "You big-headed fucking bastard."

Simon joins in the laughter at this point and realizes he

---

* Considering how many of these it is possible to commit while being English, it's a wonder we don't have a word of our own.

has to qualify his boast. He decides to take a clever route, simultaneously saying that it was the message, rather than himself, that was so popular, and making his colleagues feel the glow of a bit of regional patriotism:

"Yeah 'cos I were all like that 'Good old North East.'" The lads laugh and the crisis is averted.

Banter can be silly and inclusive or it can be barbed and meaningful. The fun kind is generally accepted, it would seem, but Barbara's research shows that some people don't respond well to banter even when it doesn't have a barbed edge. Cultural differences can cause big problems.

With forty-five employees, the company that Barbara calls "BytesBiz" in her paper was the largest of the three that she studied.* Of the three it was also the one that had the most gentle banter. It still wasn't gentle enough for new employee Brenda. At the very beginning of her time there she took a colleague, Cathy, to task for telling a customer he was "being a wanker today." Apparently the customer and Cathy both thought this was hilarious—it was part of their ongoing, jokey-friendly relationship—but Brenda was furious. She gave Cathy a telling off for behaving in a way that, to Brenda at least, was completely unacceptable.

Cathy, who'd been with the company for three years at that point, took umbrage at being reprimanded by a newcomer. The incident is so well preserved in BytesBiz employee legend that five different interviewees independently told Barbara how pissed off (or "perturbed" as Mr. O'Connor would prefer) they'd been that Brenda had dared to criticize their way of interacting with each other. In the end, Brenda left the company. She described her time there

---

* The companies as well as the individuals have pseudonyms.

as being "a square peg in a round hole" and described the company as a "zoo." Even though Brenda was never the butt of the jokes, she found the whole atmosphere toxic. As Janet Holmes discovered, resisting the swearing culture of an organization can really mark you as an outsider.

In her research, Barbara Plester also found that insults and humor only circulated in earnest among the general staff—managers didn't usually join in with that sort of jocular abuse. The problem, explains Barbara, is that if you have a superior insulting or swearing at a subordinate, that's where the trouble really starts. They could make self-deprecating jokes, but the danger of making a joke that seriously insults someone (or worse, that nobody finds funny) was seen as too big a risk to take.

A study from the United States in 1982 showed the opposite effect, however. Managers made jokes about their employees with impunity but employees were much more circumspect. I asked Barbara whether she thought this difference was due to the distance between Antipodean and North American management styles, or because times have changed over the last thirty years.

"I think maybe things have changed. Some of the managers I spoke to were so careful they were scared to abuse their power with their employees so they encouraged their employees to joke but yeah . . ." She trails off. Managers, it seems, don't have the same liberty as their employees.

Nevertheless, managers do seem to have an influence on the boisterousness of a workplace, even if they aren't setting a direct example. One of the three companies was run by two men who were, Barbara says, much gentler and quieter in their own interactions than the managers of the other

two companies. People still took the piss, but they did it in a way that wasn't as profanity-strewn or raucous as the other two companies.

Getting Your Damn Point Across
—Swearing as Rhetoric

So swearing, when used reciprocally and in good fun, might help to bond a team, but does swearing really help you get things done? In their paper "Indecent Influence," Dr. Cory Scherer and Dr. Brad Sagarin from the Northern Illinois University decided to test the use of a single, mild swear word on the way in which a message is received.[6]

Scherer and Sagarin knew from previous research carried out in the 1990s that—at least when we hear a message we disagree with—we tend to react with disgust and reject both the messenger and the message. They wondered whether the same effect held true for a message that the audience was sympathetic to. They showed a video of a speech to eighty-eight of their undergraduate students individually. The speech was about lowering tuition fees at a neighboring university. What the students didn't know was that each person saw one of three different versions of the speech at random. One version included a mild swear word ("Lowering of tuition is not only a great idea, but damn it, also the most reasonable one"), one opened with it ("Damn it, I think lowering tuition is a great idea"), and one had no swearing at all. The actor delivering the speech did his best to keep every other part of his delivery the same between speeches.

The students who saw the video with the swearing at

the beginning or in the middle rated the speaker as more intense, but no less credible, than the ones who saw the speech with no swearing. What's more, the students who saw the videos with the swearing were significantly more in favor of lowering tuition fees after seeing the video than the students who didn't hear the swear word.

Research published in January 2017 suggests that the students might have been making a snap decision about the speaker's honesty: swearing is positively correlated with honesty.[7] Dr. David Stillwell, a lecturer in Big Data Analytics at the University of Cambridge, together with colleagues from the Netherlands, the United States, and Hong Kong, studied the relationship between swearing and lying in two ways. Firstly, they asked 276 people to list all the swear words that they knew and used, and also gave them a test called "the Lie scale." The Lie scale is a 12-question form that asks for yes or no answers to questions like, "If you say you will do something, do you always keep your promise no matter how inconvenient it might be?" and "Are all your habits good and desirable ones?"

According to Dr. Stillwell and colleagues, "In these examples, positive answers are considered unrealistic and therefore most likely a lie." Honest people tend to score quite poorly on the questionnaire, because, let's face it, who among us has only good and desirable habits? The team found that people who said they swore less, and who claimed to know fewer swear words, were also likely to be either surprising paragons of virtue or pretty big fibbers.

But maybe these participants were just lying about their swearing use as well as their saintly habits? To rule

that out, the team turned to social media. They studied over 73,000 Facebook profiles and found that people who swore more on their feeds also tended to use linguistic constructions that indicate honesty.[8] For example, research shows that we use "I" and "me" far less often when we're being deceptive. We also prefer simpler words, because lying takes up valuable mental processing. Dr. Stillwell and the team compared the rates of profanity on Facebook profiles against the likelihood of a person using dissembling language. The greater the frequency of swearing, the less likely it was that a person would use dissembling language.

However, if you ask people what they think about swearing, they tend to insist that it diminishes the speaker's credibility and persuasiveness, especially if the speaker is a woman. Doctors Eric Rassin and Simone van der Heijden at Erasmus University in the Netherlands quizzed seventy-six people to see whether they thought someone was more or less likely to be telling the truth if they swore, without giving them examples. Of those that expressed a preference, more than twice as many said that swearing makes people sound unreliable.[9]

I asked James Saunders, solicitor managing director at Saunders Law Limited, whether this was an effect he'd come across in his line of work:

> A lot of people who later come to be tried are quite grumpy at the point of arrest. It's always a matter of some amusement for a well-educated counsel to say, "I put it to you, Fred Jones, that you said, 'You're a fucking cunt . . .'" In my experience, the police don't mind that

being read out in court—in general they think it works
to the disadvantage of the defendant.

But are they right? Rassin and van der Heijden followed
up their questions about swearing and reliability with some
examples. They crafted testimonies from "witnesses"
and "defendants" in an imaginary robbery. They gave the
following statement to thirty-five women in their early
twenties and asked them to rate it from 1 (unbelievable) to
10 (convincing):

> No, God damn it. As I have stated ten times, I have
> nothing to do with that. What is this all about? I have
> been here in this shitty room for almost two hours
> now. I want to go home, or I want to be allowed to talk
> to my attorney. What a fucking mess.

They then gave a different group of women an almost
identical statement:

> No. As I have stated ten times, I have nothing to do
> with that. What is this all about? I have been here in this
> room for almost two hours now. I want to go home, or I
> want to be allowed to talk to my attorney. What a mess.

The denial that contained the swearing was rated as significantly more believable than the denial without swearing. Is
this the intensifier effect that Professor Holmes talked about?
Or is it simply because we imagine ourselves also swearing up
a storm if we were ever wrongfully arrested? The researchers
tried again with statements from the imaginary victim of the
crime, one with swearing as shown below, and one without.

That asshole pulled my bag out of my hands, and ran away. He even dragged me several meters, because I would not let go of my bag. God damn it. And who is going to pay for all that? I suggest that Mister Dirtbag does.

This time a group of fifty-four college-age men and women rated the statements and again the statement with swearing came out as being more strongly believed. How does this square with the experience of James Saunders, that swearing in a courtroom doesn't tend to do anyone any favors? When I read James these examples, he said they sounded nothing like the kinds of testimony that a jury would actually hear. For a start they are too fluent.

"What you got to realize is that people talk rubbish most of the time. Just listen to tape recordings; the most commonly used word is 'uh.' The statements in the study sound very artificial to me." It turns out that the kind of swearing that college professors invent in the safety of their own lab is very different from what you or I might say in the stressful environment of a police interview room.

I ask James whether he'd ever encourage a client to let a little artful swearing slip into their indignant protestations of innocence. It turns out that watching courtroom dramas is poor preparation for a meeting with someone at the top of the legal profession as he puts me right in no uncertain terms:

I certainly don't allow my clients to swear when giving evidence in court. Judges and magistrates have to live in the real world where words are used that they'd hope

not to hear in their own living room, but a courtroom is an artificial environment. I advise my clients that yes, it's an artificial environment but they should respect it on its own terms: wear trousers with a crease in and a shirt and tie, for example.

As James patiently points out, however, trials cost a lot of money and in reality judges tend not to call them to a halt for a bit of bad language: "If you use bad language you get ticked off by the judge: 'Please moderate your language, Mr. Smith.'"

Going too far the other way can be worse:

There's a danger if people put on airs and graces—this is something that happens particularly with young coppers who try to come across like butter wouldn't melt in their mouths. But if [what you say] sounds a bit false then that's not a good impression to make on a judge or a jury. If you're caught out once, there's a high probability that they won't believe another word you say.

That's not to say that members of the legal profession have never reciprocated. When a repeat offender told Her Honor Judge Patricia Lynch QC, "You're a cunt and I'm not," she replied, "Well, you're a bit of a cunt yourself. Being offensive to me doesn't make things better at all." After spending nearly five months deliberating the case, the Judicial Conduct Investigations Office finally decided that Ms. Lynch wouldn't face disciplinary action, given that she had made an unreserved apology and promised

never to do it again. Which I think is a shame because, to me at least, her response felt like a bit of honesty among all the ritual.

Mr. Saunders doesn't think that swearing in court will catch on, no matter what we learn from studies within the narrow confines of the psychology lab. When it comes to swearing, James thinks we should focus on the message and not the manner: "I'm not convinced that swearing makes anything more or less believable. Everybody knows some people swear—[whose] every third word is 'fucking.' People don't *automatically* think, 'I don't believe him,' but I think it's a shame that people don't focus more on content rather than expletives."

James has a point. We tend to be emotionally affected by swearing—in fact, later chapters will show that our brains are wired so that we can't *avoid* being emotionally affected by it. But quite often swearing is a coded message; it either tells us that something is a joke, or warns us that we've overstepped a mark. As we'll see in the next chapter, this is probably something we've been doing since we first developed speech.

This emotional impact might explain why Ginette, the Power Rangers' team leader, was so extremely successful as a manager. She knew that a good dose of swearing can keep people on their toes or pull them back into line. She knew that when her team swore she shouldn't take it personally. She also knew when not to swear; her individual discussions with her team showed that she was an ally and she never used swearing to whinge about her team to other managers. We know that fluency in swearing is positively correlated with fluency overall: contrary to popular

belief, people who can confidently use the widest range of vocabulary also tend to know (and deploy) the most swear words.[10] Ginette is a brilliant example of this: she is fluent in three registers: English, Samoan, and swearing. If we want to be successful when attempting persuasion or bonding we should all learn to be as adept at swearing as she is.

* * * * 5 * *

"You damn dirty ape."
(Other) Primates that Swear

When did swearing begin? Were swear words part of our first vocabulary or was bad language a later invention, an expansion pack to this ability called language? Sadly, we can't go back to observe our early ancestors to see how swearing developed. Prehistoric cultures don't leave written records and, by the time writing comes along, swearing always seems to be well established already.

It's my hypothesis that swearing started early, that it was one of the things that motivated us to develop language in the first place. In fact, I don't think we would have made it as the world's most populous primate if we hadn't learned to swear. As we've seen, swearing helps us deal better with our pain and frustration, it helps to build tighter social groups, and it's a good sign that we might be about to snap, which means that it forestalls violence. Without swearing, we'd have to resort to the biting, gouging, and shit flinging that our other primate cousins use to keep their societies in check.

If we can't observe the development of swearing directly, what we need is a society with brains and social structures somewhat like our own, but that are only just beginning to use language. Thankfully, at least one example does exist, in the shape of the chimpanzees who have been taught to use sign language over the years.

Professor Roger Fouts, founder of the Chimpanzee and Human Communication Institute in Washington state, has spent his career adopting chimpanzees and studying their behavior. He taught an extended family of chimpanzees to use sign language, and watched as they passed that language on to their children in turn.

It was this extended family of apes that first convinced me that chimpanzees can do more than simply communicate; they spontaneously learned to swear. These apes were taught language (and toilet trained) by Professor Fouts. In the process of picking up both language and taboos about bodily functions, the sign they used for excreta took on a special power. Like the human swear word "shit," the sign DIRTY and the idea it conveyed became taboo. At the same time, DIRTY became a sign that the chimpanzees used emotionally and figuratively, also like the way you or I might use "shit." If Roger made them angry they would call him "Dirty Roger," the way we might say, "Roger, you shit." Unlike their wild cousins, these chimpanzees would throw the notion of excrement instead of throwing the stuff itself.

But it can be hard to convince people that chimpanzees can communicate with humans, let alone swear. And that's partly because of one of the most famous chimpanzee experiments of all time.

Project Nim

When people think of chimpanzees who have learned sign language, they tend to think of Nim Chimpsky. He's largely famous because of the 2011 film based on his experience, and in his short life he was much publicized by Herbert Terrace at Columbia University, the scientist who studied him. Project Nim was designed to be as rigorous, as clinical, and as measured as possible, which might well have been its downfall.

Behaviorism took psychiatry out of the wishy-washy world of Freud and Jung and put it on the same modern, abstract, quantitatively sound lines as physics or chemistry. When Terrace designed Project Nim in 1974 he was determined that the entire experiment would be a clean break from the unstructured chimpanzee fostering experiments that had been taking place for almost a hundred years by that point.

The advantage of these sorts of experiments was obvious: with a stopwatch and a clipboard and an endless supply of both rats and patience, it's possible to keep measuring performance. The conditions (and the rats) could be varied systematically by adding an extra turn in the maze here or an extra electric shock there. Research teams could look at particular variables—the size of reward, for example, or the age of the rat—and draw clear, statistically valid conclusions about the effect of each individual variable on how easily the task can be learned.

To be fair, some of those historical experiments had been pretty naive, treating chimpanzees like hairy human babies. This all began in the 1880s when George Romanes, a naturalist working at London Zoo, noticed the close relationship

between one of the zoo's chimps and her keeper. The chimpanzee was able to follow quite detailed instructions from her keeper and, Romanes wrote, "she resembles a child shortly before it begins to talk."[1]

This claim—that chimpanzees and humans weren't so different after all—outraged many. The very notion seemed to back up Darwin's claim that humans are just one member of a diverse family of species, rather than being somehow special among all living things. Nevertheless, in the 1920s, psychologist and primatologist Robert Yerkes fostered Chim and Panzee. He worked for months to get them to talk before finally admitting defeat. Chimpanzees' brains might be sophisticated enough for communication but they just don't have the same vocal apparatus as we do.[2] "Perhaps they can be taught to use their fingers, somewhat as does the deaf and dumb person," he eventually concluded.

Many scientists stressed the importance of a humanlike environment if chimpanzees were to learn that most human of skills, language. In the 1930s, in a rather daring experiment, married couple Winthrop and Luella Kellogg decided to raise a chimpanzee called Gua in their New York home as if she were a child. "If [a chimpanzee] is led about by means of a collar and chain, or if it is fed from a plate upon the floor, then these things must surely develop responses which are different from those of a human. A child itself, if similarly treated, would most certainly acquire some genuinely unchildlike reactions," they wrote.[3] They brought Gua home when she was seven months old and their firstborn child, Donald, was just ten months old. Sadly the experiment was curtailed after just nine months, when little Donald started imitating Gua as much as Gua imitated him.

That all changed with Project Nim. Herbert Terrace was determined to be far more rigorous than these cosy "chimp in the family" experiments. There would be no treating an ape like a baby in his lab; in Project Nim the researchers weren't allowed to treat their charge like a child, or even to comfort him if he cried out at night. Nim was an abstract behavioral subject, to be treated with as much detachment as a rat in a maze. Terrace wanted to quantify exactly how much reward would be required to teach a chimpanzee grammar. The experiment didn't lead to elaborate sentence structures or an understanding of tenses. Instead, it mainly led to a tragically maladjusted chimpanzee who had been taken from his own kind, but never made to feel part of the human world.

All of Nim's learning was carried out through a rather cold and formal process of demonstrating a sign to be copied and rewarding him only if he got the correct answer—that is, if he repeated the sign exactly. It was behaviorism at its most measured, and absolutely nothing like the way in which humans learn to communicate. There was no intrinsic reward for signing, no joy of communication. Nim had over sixty teachers who came and went fairly rapidly as well as around 150 other people who spent just a single session with him. Some of these people did at least try to make conversation, but most of them just grilled Nim with questions like WHAT THAT? WHO THIS?[4]

Imagine trying to teach a child to communicate in this way: offering a treat for saying a word correctly but ignoring them otherwise. Unsurprisingly, Nim didn't learn to do much except repeat signs for treats. Unlike Roger Fouts's adopted chimpanzees, Nim never learned to talk about his

life or the important things around him: he was rarely given the chance to have a real conversation.

Terrace noticed that Nim would grab for a treat after he'd performed a sign, even if the sign was halfhearted or badly executed. Offering treats for signs is as behaviorist as it gets: stimulus goes in, response comes out. But that isn't the way that children learn. Instead, they witness communication that goes on around them and includes them. Herbert Terrace believed that Nim's behavior showed how unlike a child he was, when actually it's a pretty good sign of how smart—and how humanlike—chimpanzee behavior is.

Had Herbert Terrace focused more on social psychology and personality studies—something that probably seemed hopelessly soft to a man of his hard behaviorist credentials —he might have considered that human children really do behave in the same lazy and halfhearted way as Nim, just as soon as treats enter the equation. Not because they're stupid, but because they have a surprisingly smart appreciation of basic economics. Give most children crayons and paper and they'll happily draw for the fun of it; the intrinsic reward of doing something creative keeps them happy and interested. But as soon as you pay children for their art, their drawings get sloppy and less detailed. They also don't seem to enjoy the process of drawing anywhere near as much when they are offered a treat in return for each piece produced. In studies, children who know they will be rewarded for their drawings spend only about half as much time playing with crayons as those children who aren't offered a reward.[5]

In fact, the exact same behavior had already been observed in the 1960s with chimpanzees in the wild. Desmond Morris, a man with "surrealist painter" and

"children's author" alongside "world-renowned zoologist" on his CV, observed that wild chimpanzees stopped drawing for its own sake as soon as they learned that drawings earned treats. Those drawings that they could be persuaded to produce were made with less time, care, and attention. "Any old scribble would do and then it would immediately hold out its hand for a reward. The careful attention the animal had paid previously to design, rhythm, balance and composition was gone and the worst kind of commercial art was born."[6] But Morris, with his eclectic career and his messy and naturalistic observations of chimpanzees in the wild, was definitely not of the rigorous, "objective," behaviorist school and his ideas never made it into the design of Project Nim.

That's a tragedy, both for us and for Nim. Nim's life was definitely one of a research animal rather than a member of a society, let alone a family. Terrace's experiments were designed the way they were with good intentions: if you're going to capture and train a wild animal, then the data you collect should at least be indisputable, right? But Nim never got the chance to be himself, and we never really did learn what he might have been capable of.

Living with Humans: Project Washoe

Lab trials like Project Nim might be rigorous, but they're also limited. Even if there were no ethical concerns about keeping a conscious and self-aware creature in such spartan conditions, counting stimulus-response pairs can never answer the most interesting questions: Are chimpanzees like us? Can they communicate? What do they think about?

If we want to know how chimpanzees fare in humanlike tasks, it's necessary to give them as humanlike an upbringing as possible. Chimpanzees that are raised in cages, with little stimulation and no social structure, are never going to learn to communicate. What the hell would be the incentive, when making signs gets you treats but trying to have a conversation gets you nowhere?

What's worse is that these lab environments harm chimpanzee intelligence. We now know that chimpanzees raised in sterile laboratory conditions actually begin to lose intelligence over time. Younger chimpanzees who have only recently been captured from the wild tend to outperform their older fellows. Without a rich and varied environment, the rich and varied intelligence of chimpanzees withers and fades.

With this in mind, Beatrix and Allen Gardner, researchers at the University of Nevada, decided to run an intensive and detailed study of the kind that Robert Yerkes and the Kelloggs had first attempted. They gave over the yard of their home to a succession of young chimpanzees who experienced something like a suburban American family life. The Gardners' first fosterling, Washoe, was captured at around ten months old in Africa and arrived at the Gardners' home in summer 1966. She was given a trailer of her own in the back of the Gardner's one-story faculty house in suburban Reno and over the next four years she learned to drink from a cup, eat with a knife and fork, dress and undress herself, and use the toilet.[7] She also liked to play with dolls, bathing them and feeding them, and learned to be pretty handy with screwdrivers and hammers. Washoe wasn't a pet; she was raised as much like a child as possible at all times.

The Gardners then fostered four newborn chimpanzees, Moja, Pili, Tatu, and Dar, between 1972 and 1976 and taught them in the same way they had taught Washoe: the same way you or I would teach a child. They called the chimpanzees' attention to things around them—dogs, people, food, toys—and showed them the signs. The Gardners and their team of "nannies"—research assistants on the project— asked the baby chimpanzees questions and responded to their requests. In short, they "taught" language by having lots of conversations.

But behaviorism hadn't gone away. One of the country's leading behaviorists, B. F. Skinner himself, criticized the Gardners' approach as being too woolly. After seeing a documentary about the project in 1974 that showed the research assistants acting as friends and caregivers to the cross-fostered chimpanzees, Skinner felt moved to write to the couple to tell them just where they were going wrong: "I was quite unhappy about your new recruits—the young people working with the new chimps. They were not arranging effective contingencies of reinforcement. Indeed they were treating the subjects very much like spoiled children. A first course in behavior modification might save a good deal of time and lead more directly to results."

But of course, treating the chimpanzees like humans was exactly the point of the Gardners' research. Human offspring were the only primates known to learn language, so the same learning methods should be used to teach language to a chimp. "Effective contingencies of reinforcement" have nothing to do with babies' first words! Human children learn to talk because it's the way our families and our societies work. And human children learn to swear because, in

human society, we experience conflict and friction. Chimpanzee society is also rife with conflict and friction and—as soon as they obtain the gift of language—chimpanzees spontaneously invent swearing.

Why Teach Chimpanzees to Sign?

The "why chimps?" part of this question has a simple answer. The similarities between them and us are remarkable, especially when it comes to their ability to learn.

Chimpanzees are surprisingly close cousins of ours. The evolutionary record shows that they are more closely related to us than they are to either gorillas or orangutans.[8] They are also similar to humans in that they have a long childhood. Most animals are born with almost all the behaviors they need in order to survive—we call these "precocial" species—but chimpanzees and humans are both "altricial," which means that we spend a very large proportion of our lives totally dependent on adults before we are able to survive on our own. This might not seem like a huge advantage, especially not from the point of view of a helpless infant, but a lengthy childhood is one of the reasons why chimpanzees and humans have such adaptable intelligences.

Whatever environment, whatever challenges we happen to be born into, we learn from those around us. This makes both humans and chimpanzees capable of assimilating new information for many years. We do this most rapidly in childhood, but both humans and chimpanzees continue to learn throughout their lives—which for chimpanzees can be as long as sixty years in captivity, or forty in the wild. A

helpless childhood is a small price to pay for a lifetime of learning.

But if chimpanzees are so similar to us, why did the Gardners teach them to sign rather than speak? That's because—as Robert Yerkes suspected—they're physically unable to copy our way of talking. Despite their ability to vocalize, chimpanzees just aren't equipped to make the same range of sounds that we use in speech. Chimpanzees have a thin tongue and high larynx, making it hard to make human sounds. What's more, chimpanzees aren't particularly vocal in the wild. According to Roger Fouts, who founded Project Washoe, chimpanzees are keen to imitate our actions but show very little interest in the sounds we make.[9] They save vocalization for purely emotional signals instead.

In the wild they make very little noise at all, and use gesture to communicate. According to Allen Gardner, "Chimpanzees are silent most of the time. A group of ten wild chimpanzees of assorted ages and sexes feeding peacefully in a fig tree makes so little sound that an inexperienced observer passing below can fail to detect them."[10]

The image we have of chimpanzees chattering to each other, which comes from films and TV shows, is nothing more than Hollywood mythmaking. In fact, the sounds we think of as chimpanzee talk are, in reality, miserable distress calls, usually recorded off-screen while some unfortunate production assistant harasses the even more unfortunate ape. According to Allen Gardner, anyone familiar with real chimpanzee vocalizations finds listening to these sounds as disturbing and distressing as hearing human cries for help: "It is easy to imagine the unpleasant scenes that evoked these high-pitched, nattering cries," he says.[11]

Primatologists know that chimpanzees vocalize only when they're in a state of extreme emotion—they make pant-grunts and pant-barks when threatened, and screams or whimpers when frightened or distressed. Allen and Beatrix Gardner wanted to understand when the chimpanzees in their care would use vocalizations and when they would stick to signing, so they ran an experiment. They would tell the chimpanzees Tatu and Dar about things that were exciting (such as going out or being given ice cream) or upsetting (like having a favorite toy taken away). They noticed something surprisingly human: when the emotional event being talked about was a long time away in either the future or the past, Tatu and Dar would sign with the humans around them. However, when the event was actually taking place they would use emotional cries.[12] Think of a human analogy—imagine watching a soccer match. Your team comes close to scoring and you cheer. The ball is deflected by the goalie and you groan. But when you talk about these things afterward, you might swear and rant and rave but we tend to use our words rather than our cries to talk about the game's events.

The Gardners and their team committed to only ever use sign language rather than spoken language around their fosterlings. Earlier in her career, Beatrix Gardner had studied the observation of animal behavior under Nobel Prize winner Niko Tinbergen and, from these observations, she knew that gesture was far more important than vocalization to chimpanzees. And so, from the very start of her life with the Gardners, Washoe was surrounded by human "friends" who didn't speak, but who exclusively talked in sign language to her and to each other: every single member

of the project team took a vow of silence around Washoe: she never heard her human companions speaking.[13]

Why the embargo on human speech? We know that human children also pick up the rules of conversation not just from being spoken to, but from watching other, older humans talk to each other. That meant that the team would have to have all their conversations in a way that the chimpanzees could potentially understand. In order that the chimpanzees wouldn't think of sign language as something unusual, the team took drastic lengths to avoid other humans who might try to speak to them. Outings for treats at Dairy Queen and McDonald's were carried out in an atmosphere of stealth: Washoe and one adult would stay in the car at a secluded parking lot while another person went to buy the food. If they were spotted, the person in the car would drive off with Washoe to prevent any interaction with curious onlookers, leaving the other passenger stranded with the takeaway until the coast was clear.

Over time it became apparent that the humans in the lab tended not to be the ones to initiate conversations—they were too busy recording observations, writing reports, and running the experiment. Washoe and the other chimpanzees learned to get their human companions' attention by making sounds, but they immediately switched to signing once they had done so.[14]

But the chimpanzees mainly talked to each other, and to themselves. Unlike in Project Nim, signing wasn't just a way of getting something from the humans. "Washoe, Moja, Pili, Tatu, and Dar signed to friends and to strangers. They signed to themselves and to each other, to dogs, cats, toys, tools, even to trees. We did not have to

tempt them with treats . . . most of the signing was initiated by the young chimpanzees," reported the Gardners. They even witnessed the chimpanzees "reading aloud." Washoe would correct herself when signing to herself while leafing through magazines. Allen Gardner recalls Washoe looking at an advertisement for a drink. THAT FOOD, she signed to herself, then examined her hand before changing the sign to THAT DRINK.[15] Washoe would also sign QUIET to herself when sneaking somewhere she shouldn't go, and sign-talk to her dolls the way a human child would speak to theirs.

After about ten months, Washoe was signing short sentences like GIMME SWEET and YOU ME GO OUT HURRY. "Washoe was thinking abstractly like a human child," says Roger Fouts, "but she was also *communicating* like a human child. She wasn't just learning symbols, she was using them to share her feelings, to control her backyard world and to get her way in every imaginable situation."[16]

Washoe also had a good understanding of the difference between what she knew and what others could know. In human children, this ability develops at around the age of four, when we figure out that just because we saw something happen, that doesn't mean everyone else knows about it. Washoe, too, understood that she could see and know things that her human friends did not. From her treetop vantage point she could see people arriving before any of her human friends and would "announce" them as they arrived at the Gardners' home.

As with any small child, this fluency with signing and the ability to understand conversations was a two-edged sword: for example, like many parents, the humans on the team took to spelling out "B-A-T-H" rather than using the actual

sign because of the way that the chimpanzees would react if they knew they were about to be bathed.[17]

The question of whether or not chimpanzees can swear, of course, demands a positive answer to the question "Can chimpanzees communicate?" Communication is necessary (though not sufficient) for swearing to take place. Were Washoe and her family's signs really language, or just another, elaborate way of begging for treats? I firmly believe that they are true language. For a start, while some of the signs were fairly obvious ones for a hungry chimpanzee to learn, much of their vocabulary had nothing to do with treats. While they learned twenty-nine signs to do with food and drink, they also readily learned the signs for things like WRISTWATCH, PHONE, STAMP, KEY, and VACCUUM CLEANER.[18] Anyone who has both a smartphone and a small child will recognize the fascination that adult objects held for the chimpanzees.

What's more, the chimpanzees started to create compound words as soon as they had eight or ten signs in their vocabularies, even though they had never explicitly been taught to do so.[19] In the same way that my niece, Romilly, spontaneously invented the tuple "eye bubbles" for spectacles, Washoe invented OPEN FOOD DRINK for refrigerator, CANDY DRINK for watermelon and CRY HURT FOOD for radish. Moja spontaneously described a cigarette lighter as METAL HOT and a thermos flask as METAL CUP DRINK COFFEE, while Alka Seltzer was LISTEN DRINK. They could also answer questions about the features of the things and people around them. The researchers would ask questions to see whether the chimpanzees understood the relationship between signs and the real things they represent, much like we might ask a small child, "What noise does the duck

make?" or, "What animal goes moo?" and there are many examples of conversations about the redness of Washoe's favorite boots, or to whom the METAL HOT belongs, to show that they could understand the relationships between signs and objects, the signifier and the signified.

The chimpanzees were, I think it's safe to argue, linguistically very talented indeed. They could understand signs, invent new meanings, and talk about their world. But I want to go one further and make the argument that these chimpanzees didn't just invent clever new names for watermelon and Alka Seltzer, or new ways of being emphatic. As soon as they mastered language, they invented their own swearing. But first, they needed one other skill.

When You Live with Humans, DIRTY Is Taboo

I don't think Washoe would have learned to swear if the people with whom she lived hadn't decided to potty train her. That said, I can't imagine anyone being brave enough to test my conjecture; living with a chimpanzee long enough to teach them sign language without potty training them would be an interesting and malodorous challenge. Wild chimpanzees deliberately piss and shit on the human researchers who visit their turf in a grim territorial display that is impossible to ignore.[20] Knowing this, the Gardners weren't taking any chances; if Washoe was going to live in their home, they were going to have to teach her that there is an appropriate time and place to excrete—at least while she was among her human family.

Like many toddlers, Washoe didn't find potty training

particularly easy. The taboos surrounding bodily functions were already ingrained and Washoe had become "potty shy"—she would rather use her diaper than the pot.

The team realized that they needed to teach Washoe that DIRTY might be bad but DIRTY in the potty is fine. Roger Fouts began his career as a research assistant to the Gardners, and was largely responsible for that potty training. He remembers these as some of the most challenging weeks with Washoe. "This request soon became so routine that poor Washoe sometimes sat on the potty while I begged her in sign: PLEASE PLEASE TRY or PLEASE TRY MAKE MORE WATER," he recalls.

Washoe had become more modest than some humans; the young chimpanzee had internalized many of the taboos and standards of politeness of a suburban American family. Allen Gardner recalls that Washoe "seemed embarrassed when she could not find a toilet on an outing in the woods, eventually using a discarded coffee can."[21] The Gardners' chimpanzee family no longer shit in the woods, having learned that DIRTY anywhere but the potty is BAD.

Among Washoe and the other chimpanzees raised by the Gardners and their team, the DIRTY sign was consistently used by chimpanzees and humans alike for feces, dirty clothes and shoes, and for bodily functions. Double use of the word DIRTY was used to intensify the meaning, either in anger or shame. DIRTY DIRTY SORRY was a phrase used to apologize for accidents, while DIRTY GOOD was the name that Washoe used for her potty.[22] This name, invented spontaneously by Washoe, shows a surprisingly nuanced understanding of the excretion taboo: pooing in a potty is necessary and acceptable, but shit out of context is shameful and wrong.

One other thing that persuades me that DIRTY is taboo to chimpanzees is that they, like us, are motivated to lie if discovered doing something shameful. The mortification of being caught out in a transgressive act prompts humans to tell some enormous whoppers. Again, anyone familiar with toddlers or young children will have seen the transparent and sometimes hilarious lies they will tell when caught red-handed, like the three-year-old I saw recently, solemnly swearing, through lips ringed with jam and sugar, that her three-month-old baby brother had eaten the last doughnut. From the data collected by Roger Fouts we know that DIRTY was something shameful enough for the chimpanzees to lie about in much the same way. Take Lucy: she was a chimpanzee who wasn't originally raised by the Gardners, but whom Roger studied in her foster home with another family.

Lucy was a show-off and didn't like it when she wasn't the center of attention. If Roger ignored her while he was speaking with her human family for too long she'd relieve herself in the middle of the living-room floor. It's not that she wasn't potty trained—if she was in a good mood she'd poo in the appropriate place. But when crossed, Lucy would stage a dirty protest. Here's one conversation she had with Roger:

Roger comes into the room—sees Lucy's little message and asks: WHAT THAT?

WHAT THAT? Lucy replies with what seems to me like artfully feigned innocence.

Roger isn't fooled, though: YOU DO KNOW. WHAT THAT?

Lucy replies, DIRTY DIRTY.

Roger asks WHOSE DIRTY DIRTY?

SUE signs Lucy. (Sue was Roger's assistant and a graduate student at the time. While I can vouch for the fact that stress makes PhD students do odd things sometimes, Roger is pretty certain that Sue wouldn't go that far.)

He tries again. IT NOT SUE. WHOSE THAT?

ROGER, replies Lucy, in what is, admittedly, a pretty desperate attempt to shift the blame.

Roger scolds her: NO. NOT MINE. WHOSE?

Lucy: LUCY DIRTY DIRTY. SORRY LUCY.

Perhaps because of the shame associated with it, DIRTY soon became an insult, used when people or other animals didn't do what Washoe wanted. This wasn't something Washoe was taught to do; she spontaneously began using DIRTY as a pejorative and as an exclamation whenever she was frustrated. When we internalize a taboo, chimpanzees and humans alike create an emotional connection with the concept. The words for taboo subjects don't just cause strong emotions; they leap to mind whenever we experience strong emotions. For example Washoe signed DIRTY ROGER when Fouts wouldn't let her out of her cage and DIRTY MONKEY at a macaque who threatened her.

In fact, MONKEY became Washoe's somewhat derogatory sign for any other primate who couldn't sign. Somewhat depressingly, it seems as though slurs are another deeply ingrained part of our language.

Although Washoe and her peers had only one sign that they used as swearing, they used it with a great deal of flexibility. In the same way that "fuck" and its variants are often hissed, shouted, or spat, DIRTY would be signed with considerable emphasis by the chimpanzees. "The DIRTY sign is the back of the wrist brought up against the underside of the

chin and sometimes the chimpanzees would make this sign so emphatically that the clacks could be heard throughout the lab."[23] You can imagine the strength of Washoe's disapproval as she bangs her wrist into her jaw, smacking her teeth together. It's an image that is both forceful and very human and reminds me of the way that the middle finger can be brandished or the fist pounded into the crook of the elbow when some utter shithead cuts you off in traffic. (Surely that's not just me?)

But, like humans, the chimps don't just rely on the shit-taboo as a means of hurling abuse and expressing anger. As every small child knows, scatological is funny; its power to shock causes a reaction that can startle us into a laugh. By the time the DIRTY taboo had been internalized it became clear that Washoe wasn't above a bit of scatological humor. She cottoned on to the fact that the human observers didn't like cleaning up her mess—initially this became a weapon to control the team: "it didn't take long for her to learn that she could manipulate us by having, or threatening to have, an accident. It must have been terrific fun to go high up in her tree and commit a simple, natural act that would cause grown-up humans to jump around in desperation on the ground below," speculates Roger.

The power of making the humans scurry around at her bidding might have been satisfying, but Washoe soon discovered that dirty jokes could be funny for their own sake. She really enjoyed piggybacks and would ride on Roger's shoulders for a treat. One day, Roger noticed that Washoe signed FUNNY to herself while snorting. "For a second," says Roger, "I couldn't figure out what was funny. Then I felt something wet and warm flowing down my back and into my pants. I never forgot the sign for FUNNY after that."

These chimpanzees, raised like children, ended up behaving like children. It's so compellingly human—the lying, the joking, and the shame that are connected with our bodily functions. The taboo becomes so powerful that it can be used to berate and control (DIRTY ROGER, DIRTY MONKEY) but it also becomes the basis of hilarious ribaldry. There's absolutely no doubt in my mind that Washoe, Moja, Pili, Tatu, Dar, and Lucy all learned to swear, as soon as they learned what a taboo was. That same emotional connection exists in chimpanzees and humans alike.

Teaching the Kids: Can Chimpanzees Pass on Language?

Roger and Washoe eventually left the Gardners' lab and set up on their own. Because Washoe showed such strong maternal instinct,* Roger decided to contact other primate research institutes to see if there were any infant chimpanzees who needed adoption. Roger brought Washoe ten-month-old Loulis and, after a slow start, they soon bonded as mother and child.

Roger and Deborah Fouts made the decision to change life for Washoe drastically at this point. In order to see if language could be learned across generations, they banned

---

* Washoe gave birth to a son, Sequoyah, while she and Roger were based at the University of Oklahoma. Sequoyah became injured and died of an infection when he was just days old. Roger describes Washoe's grief, and his fury, in Next of Kin, which changed the way I think about our relationship with other primates. I defy you to read about Washoe's time in Oklahoma without wanting to swear.

all humans from signing with Washoe whenever Loulis was around. Because Loulis and Washoe were inseparable there was, in practice, no time when Washoe could sign with her human friends anymore. For five years, Washoe had hardly any conversations with the human friends who had raised her since infancy. For five long, sad years, the chimpanzees and humans were no longer on speaking terms.[24]

In that time, Washoe taught Loulis to sign in the same way that she had been taught, modeling some signs for Loulis and shaping his hands to make others, directing his attention to objects or encouraging him to do certain things. By fifteen months, five months after being adopted by Washoe, Loulis had learned several signs and had also started to combine them into two-sign compounds like HURRY GIMME and PERSON COME. The Gardners had continued to foster infant chimpanzees throughout the 1970s, and as these chimpanzees reached adolescence they were sent to join Project Washoe. In 1979, Washoe and Loulis were joined by Moja, and in 1981 five-year-olds Tatu and Dar joined the group. All five signed back and forth both with each other and with Washoe's adopted son, reinforcing Loulis's learning and adding to his fluency.

In the time that he was exclusively signing with other chimpanzees, Loulis learned fifty-one words including BOOK, DIRTY, HUG, PLEASE, and SORRY. Washoe, too, was still learning: Moja, Tatu, and Dar had learned the new word BLANKET while living with the Gardners in Reno and Washoe soon started using it, showing that the ability to learn language persists well into adulthood for chimpanzees as well as humans.

Between 1980 and 1993 the Foutses and their chimpanzee family were based at Central Washington University.

They had left the University of Oklahoma, where the facilities for Washoe had been unsafe and unwelcoming, for Washington, which was far more supportive of the aims of Project Washoe. Although CWU was supportive of the program, a lack of funding meant that the facilities were very limited. Washoe, Loulis, Moja, Tatu, and Dar found themselves confined to a 300-square-foot home on the third floor of the university's psychology building. Having moved from the Gardners' suburban home in Reno to the prison camp in Oklahoma, Washoe and her family now lived in the equivalent of an urban apartment. They had no access to the outdoors and the only natural light they saw came from a single window. The chimpanzees lived as a family, ate raisins, and drank Kool-Aid, played games of chase and tickle with each other and with the research staff, but it was a cramped and unnatural life for even these chimpanzees.

Roger and Deborah Fouts worked tirelessly to try to change the chimpanzees' living conditions. In May 1993, after much lobbying of Washington state legislators and donations from the Friends of Washoe charity, they finally moved the chimpanzees. Their new home had over 7,000 square feet planted with grasses, bamboo, and wild plants and equipped with swings and climbing frames, fire hoses, and a vegetable garden that the chimpanzees liked to discuss almost as much as they liked to eat from.

This new space was a huge change for Loulis, who had not seen the outdoors for most of his fifteen years. At first he was nervous, and the human research team had a plan to introduce the chimpanzees to the new space in stages, letting them adjust to the move before gradually opening up the indoor and then the outdoor space. But Washoe had

watched the staff discussing the move in sign language, so she knew what was waiting for them beyond the doors. She'd been missing the outside world for over a decade. "Our plan quickly changed when a few hours after moving in, Washoe woke up and looked outside through the glass enclosures and . . . began signing OUT OUT OUT THERE," said Roger.[25]

It was here, in a place of both space and safety, where Washoe was to live out the rest of her forty-two years. In 2007 she died in her bed, surrounded by primates of two species who loved her.

Does It Matter If Chimpanzees Can Swear?

Chimpanzees like Washoe—and people like the Fouts and the Gardners—have taught us so much about our next of kin in the animal kingdom. Through a heroic half-century of living with chimpanzees, treating them as our cousins and observing how they behave, Project Washoe gives us a clearer idea than ever of what our own pre-*sapiens* origins might have been like.

There are a lot of things that convince me that Washoe and her family had self-awareness and a rich emotional life, but none more than her ability to notice when people were sad or hurt and to offer comfort. Roger tells the story of when one of his colleagues, Kat, returned to work after losing her baby. Washoe wanted to know why Kat had been away. Kat signed to Washoe: my baby died. "Washoe peered into Kat's eyes again and carefully signed CRY, touching her cheek and drawing her finger down the path a tear would make on a human," said Roger.[26]

Chimpanzees aren't primitive humans. They have their own distinct genetics and evolutionary history. But they are our closest cousins, and by studying them we have filled in some of the gaps in our own evolutionary record. They share our gifts for humor, for compassion, and for learning as well as our burdens of conflict and frustration. Washoe and her extended family have shown us that, whatever else our ancestors did, it's a pretty safe bet that they swore almost as soon as they learned to talk.

Whether you're a chimpanzee or a human being, in order to swear you need an understanding of the psychology of others, a working theory of mind, to be able to anticipate how your words are likely to make someone else feel. You also need an emotional life; without meaningful emotions there would be no swearing. You need a complex enough mind to understand social concepts like taboos; if we didn't have an idea—however vague—of a society that disapproves of some things and approves of others, we would never know shame or taboos and there would be no such thing as swearing.

And we can be thankful that our ancestors swore, too. As a safety valve and a bonding mechanism, swearing has no equal. Those proto-humans who first banded together to hunt might never have been so successful if it weren't for the gift of swearing that developed right alongside the very first speech.

It matters that chimpanzees can communicate because we finally have evidence of a non-human intelligence living here on earth. For years, in my discipline of artificial intelligence, we've been debating the ethics of how we should treat non-human intelligence if we ever manage to create

it. But non-human intelligence already exists: chimpanzees can think, feel, want, grieve, teach, fear, and feel shame and compassion. They are self-aware enough to communicate, and have a complex enough internal life to recognize taboos, which they can use to swear, and they have strong emotions that drive them to swear. In June 2013 the American National Institutes of Health said they would no longer give financial support to experiments on chimpanzees, and in June 2015 the US Fish and Wildlife Service put captive chimpanzees on the "endangered" list, meaning that anyone who wants to carry out research using chimpanzees as subjects will have to apply for a permit.

Modern medicine is based on the use of uncounted numbers of animal test subjects from rats and rabbits to apes and monkeys, but I'm profoundly relieved that we've finally recognized the intelligence that exists in chimpanzees. And nowhere is that intelligence more obvious than in their invention of swearing.

***** 6 *

No Language for a Lady: Gender and Swearing

One of the things I've learned as a woman in a male-dominated profession is that a laissez-faire attitude to swearing can go a long way. As a shortcut to being accepted as one of the boys it's more effective than learning the offside rule and easier than burping the theme song to the Muppets. But while it might make me one of the lads in the lab, swearing is likely to hold me back in the real world, while it's more likely to help my male counterparts. Whether we like it or not, research shows that we are still a lot more judgmental of women who use taboo language than we are of men.[1]

But this double standard is relatively new. Sometime around the early eighteenth century there was a significant change in culture. Male and female language began to undergo a shift that can best be described as power for men versus purity for women. Influential commentators at the time encouraged women to adopt a "clean" style of language, shunning taboo words—especially those that related to bodily functions—on

pain of social exclusion and the threat of an eternity in hell. Faced with such impressive threats, women devised increasingly elaborate euphemisms as they sought to say what had become unsayable. Men, on the other hand, were expected to master power, and that power can still be seen today. Men's verbal styles tend to be more direct than women's: women are more likely to use euphemisms ("I'm going to powder my nose") and are more likely to hedge requests with face-saving get-out clauses ("Would it be OK to . . . ?").

This shift meant that swearing became part of the male side of the lexicon. Swearing isn't just full of "impurities"; it can also be very direct, sometimes even aggressive. The power of swearing isn't limited to the verbal equivalent of beating one's chest, however. In a society where only men can speak of (and by extension learn about) taboo subjects like sex and other bodily functions, that knowledge—that power—was extended to men and denied to women.

In retaining the right to swear, men also held on to the power to express a much wider range of emotional states. As swearing in the workplace shows, it's a powerful tool for joking, bonding, and being part of a team. Jocular abuse between women was written out of the language and a more indirect—and occasionally hypocritical—form of interaction had to take its place. Those insisting that women's language should be pure managed to rip the most powerful linguistic tool out of the hands (and mouths and minds) of women for centuries.

Surely things are starting to change? Here I am, a woman writing a book about swearing. As far as I know my friends don't hold me in contempt for my obsession with taboo language; my sex has never made me feel as though I shouldn't

swear. Research shows that women are using swearing and other equally powerful forms of language more effectively than ever, but that same research shows that doing so still comes at a greater social risk for women: a man swearing is more likely to be seen as jocular or strong; women are likely to be seen as unstable or untrustworthy. To which I can only ask: where the fuck did this bullshit come from?

"Why won't you listen?" The Indirectness of Women's Speech

Wherever it comes from, it's pervasive. We all internalize the norms of politeness as we grow up, and what counts as polite varies depending on your social group, generation, and gender. One parameter is how bluntly we speak to each other. This quality, which linguists call directness, varies from culture to culture, but also between subcultures. As a woman who has worked in several male-dominated fields I've had to learn the hard way that statements starting with "Would you mind if I . . . ?," "I wonder if maybe . . . ?," or "Could we perhaps . . . ?" tend to go unheard, or at the very least unattended to. More direct colleagues would say, "We'll do it this way." (Even more direct ones: "Why the hell aren't we doing it this way?") When linguist Deborah Tannen studied male and female conversational styles she found that men are much more likely than women to state their beliefs, desires, and intentions directly. They're also more likely to interrupt with contradictions, whereas women respect "turn taking" and are more likely to interrupt only when they are agreeing with the speaker.[2]

Indirectness is largely about saving face, allowing the person you're talking with to disagree with what you're saying or not comply with what you're asking. "Is anyone else cold in here?" contains the tacit information that I might like the heating on but doesn't demand a yes or no answer to the question, "Please may I put the heating on?" Some cultures are much more indirect than others: when my husband and I were studying Japanese I seized with glee on all the different ways of saying no to a request that don't involve the mortification of actually saying 'No!" "It's difficult right now . . ." (complete with a slightly helpless trailing off at the end) was one of my favorites and I took great delight in jokingly using it around my considerably more direct husband, who has very few problems saying "no" and who finds my sense of obligation somewhat bizarre.

The indirect register in Japanese made me very comfortable indeed, and when I worked there I finally found myself having conversations that felt like I was on a par with my male colleagues. Only when studying the language more intensively when I returned home did I realize that the indirect register was the reason for my greater comfort. That's because, in British English at least, women and girls tend to be socialized into preferring a more collaborative, less competitive style of communication, while men and boys are generally rewarded for directness. In Japan, indirectness of expression is a widespread preference regardless of gender.

What does this have to do with swearing? Few speech acts are more direct—or more likely to lead to a loss of face if used inappropriately—than swearing. There is potentially a strong bias against swearing for many women, a bias that comes from the same social conditioning that sees us adopt

more indirect and conciliatory forms of speech. However, as more of us are working in previously male-dominated workplaces, and as later generations are starting to resist some of the constraints of traditional femininity, women and girls are changing their speech patterns. And nothing says "one of the boys" like swearing. But why is this so? Directness aside, why do we assume that women don't swear as much as men? And if we *don't* swear as much as men, then why not?

Femininity Means Never Having to Say "Fuck You" —Protecting the Fairer Sex from Fouler Language

Plenty of researchers have started with the assumption that women swear less than men and have then tried to explain why that might be the case. In the early twentieth century, linguists were still confidently asserting that "women will often invent innocent and euphemistic words and paraphrases while men tend to favour more coarse language."[3]

The most common explanation boils down to the same double standard that makes "slut" an insult and "stud" a compliment. Swearing in both sexes is associated with sexuality and—given that women are judged more harshly than men for their sexual adventures—women supposedly keep a lid on bad language to prevent accusations of bad behavior.[4] Swearing is also part of that direct register that we are comfortable hearing from men but that still seems aberrant coming from a woman. However, this attitude is relatively recent.

From the books and pamphlets that were popular in the 1600s, we know that attitudes to the allegedly fairer sex were starting to change sometime around then. Despite the fact

that women were the ones going through the rigors of pregnancy and childbirth (which, by definition, involve fucking, shitting, and bloody cunts) we were now expected to have pure minds, unsullied by even the notions of bodily functions or other taboos.

The most influential advocate for refinement in women's language and behavior was Richard Allestree. His position as provost of Eton College and chaplain to King Charles II gave him an extremely effective platform for his ideas. In his book *The Ladies' Calling* (1673) he insisted that women who swear begin to change sex, undergoing a "metamorphosis" that makes them "affectedly masculine." According to Allestree, this would be such an affront to God's order for the world that: "there is no noise on this side of Hell can be more amazingly odious" than "an Oath … out of a woman's [mouth]." What a claim! Not the crying of a hungry child or the groans of the sick—no, what really pisses God off is a woman saying "asshole."

Women's experience of pregnancy and childbirth once made them mystical and slightly terrifying creatures, but these new views on femininity attempted to control this power, to impose purity and innocence on womankind. "From being dangerous creatures, women had been reduced to passive angels by the end of the seventeenth century with femininity in part being defined by a purity of discourse," says Tony McEnery, professor of linguistics and the author of *Swearing in English*.[5]

At the time that Allestree was writing, women's social roles were narrowing. This was a process that he both welcomed and encouraged. In fact, Allestree recognizes only three types of women: "virgins, wives and widows." Any other

state disqualified one from society, or at least from the good graces of Allestree (and God's good graces, the two being indistinguishable in his mind). Foul language was a sure sign that a woman had knowledge of foul deeds. According to Mr. Allestree, women's swearing is a sign of a breakdown in the acceptable social order where masculine men were put on this earth to keep fragile femininity safe from the harsh knowledge of the world. Richard Allestree would have really hated my being a researcher—a woman whose job could best be described as professional curiosity: "There is none more mischievous than Curiosity, a temptation which foil'd human nature even in Paradise: and therefore sure a feeble girl ought not to trust herself with [it]," Allestree decreed.

This campaign soon spread beyond the policing of women's language. If men were still using profanities in the *presence* of women, who knows what we ladies might pick up? Jeremy Collier, one of the foremost campaigners against bad language at the turn of the eighteenth century, leveled his quill against the Restoration playwrights with their use of themes and language that were base, common, and popular. He accused these playwrights of using language that was likely to corrupt innocent women.[6] "Do women leave all the regards of decency and conscience behind them when they come to the play-house?" he wrote. "To treat the ladies with such stuff is no better than taking their money to abuse them. It supposes . . . that they are practised in the language of the stews, and pleased with the scenes of brutishness."

Some playwrights pointed out that ladies were still attending theaters with every appearance of enjoyment. Perhaps these "scenes of brutishness" were exactly what drew them in? Inconceivable! Collier concludes. What has

clearly happened is that these ladies have become so *completely* refined as to be entirely incapable of understanding the swearing and sexual humor shown on stage.

There were those who thought Collier a bit extreme, even by the standards of the day. Playwright Thomas D'Urfey described Collier as "foaming at the mouth" for his criticism of *Don Quixote*. But then, D'Urfey did write a song called "The Fart," so the two were never likely to see eye to eye.

Mind you, when Collier was agonizing over the poor, innocent ladies who were being "abused" by the bawdiness of Restoration comedies he did, of course, mean "ladies" rather than women in general. For there was altogether another class of females that fell entirely outside his overbearing concern: "are these [depictions of female desire and strong language] the tender things Mr. Dryden* says the ladies call on him for? I suppose he means the 'ladies' that are too modest to show their faces in the pit." (Today we might prefer the term "sex worker.") "It regales their lewdness, graces their character, and keeps up their spirits for their vocation," he goes on.

And so we begin to understand Collier's objections. It isn't women's delicate *ears* he's worried about, but an altogether different part of the anatomy. Dryden and these other irresponsible playwrights didn't care if their earthy language and earthier subject matter opened ladies' minds to the sort of behavior that the doxies in the pit enjoyed, or at least used to turn a modest profit. The idea that a woman could exist for whom sex was neither a matter of shame nor a profitable vocation was anathema to Collier and his adherents. So much of the way we judge women's language, and their

---

* John Dryden, playwright and poet, August 9, 1631–May 1, 1700.

behavior, still rests on the antiquated double standards of a handful of long-dead churchmen.

Eighteenth-Century Attitudes in Twenty-First-Century Lives

Even in the twenty-first century there are still commentators who rail on the "new" habit of women swearing like the menfolk. A quick Google search of "women swearing" brings up a slew of articles and comments about how off-putting some men (and some women) find a woman who swears.

While Google searches are highly anecdotal, numerous studies show that these attitudes are commonplace. From South Africa to Northern Ireland, Great Britain to the United States, women are judged more harshly than men for swearing. Teenage girls in the 1990s—my contemporaries and the daughters of second-wave feminists—considered swearing to be far less acceptable from a woman than a man,[7] even though women now account for 45 percent of all swearing in public (up from 33 percent in 1986).[8] In the United States at least, women have actually drawn closer to their male counterparts in swearing than in earnings, receiving just 43 cents of every dollar earned, while men earn the other 57 cents.

Dr. Karyn Stapleton at the University of Ulster has carried out many in-depth interviews to find out why this perceptual mismatch exists: why is it becoming more usual for women to swear, but these old attitudes still persist? "Generally, swearing is becoming more acceptable in society as a whole but there are still differences in how it's perceived coming from women and men," she told me. "Men tend to

feel comfortable swearing more freely across a wider range of contexts and with a wider range of people." The division of language into purity and power might still be responsible for women's reluctance to swear. "From numerous studies we know that women are expected to pay more attention to politeness than men. Swearing can be incredibly direct." Because of the double pressure of inviting moral censure and being judged harshly for failing to be polite, "choosing to swear is still a much higher risk for women."

The same attitudes exist on the other side of the Atlantic, too. Dr. Robert O'Neil of the University of Louisiana carried out a study in 2002 in which he showed transcripts containing swearing to both men and women. If he told the volunteers that the speaker was a woman, they consistently rated the swearing as more offensive than when they were told that the speaker was a man.[9] I asked him why he thought that might be. Gender roles are, it seems, still largely to blame—and these roles are as prescriptive for men as they are for women. "Men are expected to be aggressive, tough, self-reliant, always looking for sex, and most importantly, not effeminate. Women are evaluated first and foremost on appearance. But we also expect women to be caring and nurturing, and to be nice to everyone." Dr O'Neil also thinks that power plays an important role: "Because women and children have been seen as weaker than adult men, it has been deemed necessary to protect them from profanity, pornography, and other privileges of power."

As we saw in Chapter 2, on pain and swearing, women with cancer who swear even tend to alienate their friends over it, while men don't face the same condemnation (page 63). With social censure like this in place, it wouldn't be

surprising to discover that women bow to the pressure to swear less. But when you look at the numbers, women have been bucking this pressure and starting to swear more since at least the 1970s. We know that the differences between men and women swearing are not so clear-cut in practice.

!

If you want to study how people actually use language, the best place to start is with a corpus: a huge collection of written and spoken language that is representative of how that language is actually used. The British National Corpus (BNC) has thousands of examples of spoken and written English and is representative of how the language is used in British society. Created by the Oxford Text Archive in the early 1990s, it contains 90 million written words and 10 million spoken words of British English taken from letters, newspapers, books, school essays, and informal, unscripted conversations. If we want to understand the differences between the way language is used by different groups of people then we can mine this database and compare vocabularies.

Most words will be used identically by everyone. Some, such as "and," "it" and "the," will be used too commonly to show any difference. Others, such as "xylophone" or "platypus," will be used too rarely. Some words make for terrific indicators of social difference: think of the class and age distinctions in England that underlie the difference between "mum" and "mummy."

If we examine how the utterances and writing in the BNC differ between men and women we find some interesting patterns. While women swear almost as much as men in public (45 percent of public swearing in 2006),[10] the nature of male

and female swearing is quite different. In the BNC, "fuck" and its variants turned out to be the most significant indicator of male speech—that is: if the word "fuck" was used and you had to guess the sex of the speaker, you'd be right most of the time if you said that the speaker was a man. On the other hand, of the top twenty-five words that were characteristic of women's speech, not a single one was a swear word.

Does that absence of swear words in the female lexicon mean that I'm an aberration? Surely not, if women account for nearly half of all publicly recorded swearing. So what is going on? It turns out that the type and variety of women's swearing is very different from that of men's swearing. In the Lancaster Corpus of Abuse (a database that is a lot like the BNC but is restricted to swearing), Professor Tony McEnery found that British women were just as likely to swear as men.[11] The most significant difference is that women tended to use milder swear words ("god," "bloody," "pig," "hell," "bugger") than their male counterparts.

But that seems to be changing. In a brand-new study, set to be published in 2018, Professor McEnery analyzed about 10 million words of recorded speech, collected from 376 volunteers. Women's use of "fuck" and its variants has increased fivefold since the 1990s, whereas men's use has decreased. If you're interested in the exact scores, women now use "fuck" and its variants 546 times per million words, while men use them only 540 times per million.[12]

In the United States, gender differences still appear to prevail, although that might simply be an artifact of exactly who is asking the questions. Professors Lee Ann Bailey and Lenora Timm from the University of California, Davis, discovered that their male students swore more and used

stronger swear words than their female students. However, they also found that people have a strong tendency to curb their bad language in front of parents, children, and the opposite sex.[13] Might women appear to be using milder language because the majority of data collection carried out until the 1980s was done by male professors?

In order to test this hypothesis, Dr. Barbara Risch of New Mexico Highlands University carried out a study of the insulting swear words that women used for men.[14] Dr. Risch asked her female students to stay behind after class and ensured that all of the interviewers carrying out the study were also female. She asked the young women the following question: "There are many terms that men use to refer to women which women consider derogatory or sexist (*broad, chick, cunt, piece of ass*, etc.). Can you think of any similar terms or phrases that you or your friends use to refer to males?"

After some initial reticence, the women in her class—all university undergraduates—let rip with an impressive array of swearing that Dr. Risch organised into categories such as animal ("bitch," "pig," "dog"), birth ("bastard," "son of a bitch"), head ("dickhead," "shithead"), dick ("dick," "prick," "cocksucker"), and others ("jerkoff"—sadly "wanker" hadn't yet crossed the Atlantic).

With no men around, they were much more comfortable swearing. "Do women swear?" gets a different answer depending on who's asking, it seems. While Dr. Risch's study doesn't prove that women have been hiding their swearing from male sociology professors down the years, it did open up the possibility that women's swearing had been systematically underreported and that "Why don't women swear?" is actually the wrong question to ask altogether.

No Shit, Sherlock: Women Swear
Whether You Like It or Not

It isn't just American undergraduate women who are prepared to swear in front of female (but not necessarily in front of male) instructors. A slightly later study in South Africa replicated this result.[15] Dr. Vivian de Klerk at Rhodes University in Grahamstown, Eastern Cape, showed that teenage girls were just as likely to know and use bad language as boys of the same age and background. In fact, these teenage girls have an impressive array of words for boys, both ugly and good-looking: my personal favorite is the phrase "ovary overflow" for an attractive man.

In the UK too, women have been using swearing for self-expression. In the early 1990s, Dr. Susan Hughes of the University of Salford was studying the language used by women who attended a family center in Ordsall, Greater Manchester, a deprived, inner-city area of the north of England.[16] Here, male unemployment is common. Female employment is usually in menial jobs such as cleaning or factory work. The women have tiring and challenging lives as both the main breadwinners and heads of the family.

Tasked with keeping their families together in the face of poverty, unemployment, and social blight, they adopted matriarchal personas. These women needed to be obeyed and, perhaps, even a little bit feared. By using strong language these women earned respect, both from each other and from the men they taught to swear. When faced with a choice between the language of purity and the language of power, collectively they decided that purity could go and fuck itself.

Dr. Hughes was volunteering at the family center to help with a literacy program. She noticed that these women didn't speak the way the scientific literature said that they should. All of the existing literature suggested that women would be more likely to use polite forms of speech than they were to swear, particularly in their roles as mothers. Instead, she found that the women were proud of their swearing, and of passing on the skill. "It's not swearing to us. It's part of our everyday talking," said one woman at the center. "We've taught men to swear. Foreigners what's come in the pub," said another proudly.

Among these women, swearing isn't being used to shock anyone. Dr. Hughes noticed that the women didn't much care who did or did not hear, nor did they care about the effect on the listener. However, the women did differentiate between swearing as an insult and swearing "just because." "To call a child 'a little bastard' or 'a little twat' is at times almost like an endearment," she observed, although the same words were used when telling children off.

However, some swearing was off limits, or at least frowned upon. The women at the community center tended to be more religious than the general population. While they were comfortable with "bastard," "cunt," and "shit," most of them said they would never use "Jesus," "God," or "Christ."

The women in the Ordsall community center were dealing with extremely stressful circumstances. It might well be that their profanity was another example of swearing as a response to social pain. But their cursing was mainly friendly and jocular, especially among themselves and toward their children. To me, it seems to be propositional

and strategic, chosen for the effect of shrugging off restrictive gender norms and reclaiming some power.

Something happened in the twentieth century that helped to take the lid off women's swearing. In the 1970s, the women's liberation movement encouraged women to seek the same fulfillment in the workplace and in society in general as their male counterparts. Along with this social shift came a change in women's language. One of the earliest studies to show this came from Professors Marion Oliver and Joan Rubin.[17] They discovered that the greatest predictor of whether or not a woman swore was whether or not she was married: women who had rejected the status of wife and homemaker were more likely to feel comfortable swearing. For some participants in the study, there seems to have been an element of protesting too much: those women who wanted to be liberated but were still working hard at becoming so were more likely to swear than the women who already felt fully liberated. The earlier mentioned study by Professors Bailey and Timm (page 157) was carried out around the same time. In their research, they found that women in their early thirties were marginally more likely to use slightly more strong swear words than their male counterparts. Perhaps these were women who didn't feel confident in their liberated status and were overcompensating.

The willingness to swear—and even a certain pride in doing so—had spread to the UK by the turn of the twenty-first century. A study of teenagers on MySpace from 2003 by Professor Mike Thelwall of the University of Wolverhampton found that there was no significant difference between male and female British teenagers overall, but that boys were slightly more likely to use very strong (e.g. "cunt") and mild (e.g. "tosser") swearing, while girls were more likely to use strong swearing (e.g. "fuck,"

"asshole").[18] What's more, British teenagers were much more sweary than their American peers, to my immense patriotic pride. One of the young women wrote on her profile: "I am foul mouthed but if you're gonna get any man to listen to you these days you have to talk like them (motherfucker!)."

She makes an important point. Most swearing is neither aggressive nor an anguished explosion. It's something we choose to do for a deliberate effect. Ironically, what began as a move to restrict women's power has actually resulted in a kind of backhanded prestige. By refusing to accept restrictions on their language, these women are making a deliberate statement: "Take me seriously (motherfucker!)."

When women swear they tend to do so for the same reasons that men do. But discovering that women do, in fact, swear has perplexed some researchers. While there hasn't been a study to look specifically at the reasons why men swear—perhaps because it's seen as the default, normal, not in need of explaining—women's swearing seems to have demanded an explanation.

Why Do Women Swear?

Dutch linguists Eric Rassin and Peter Muris, perhaps inspired by Sigmund Freud's "unanswered question,"* published a paper with the title "Why Do Women Swear?"[19]

---

* "The great question that has never been answered, and which I have not yet been able to answer, despite my thirty years of research into the feminine soul, is 'What does a woman want?'" From *Sigmund Freud: Life and Work* by Ernest Jones (Basic Books, 1953).

Professors Rassin and Muris don't tell us whether they were surprised to discover that their female students happily confessed to using, in decreasing order of frequency, the words "shit," "*kut*" (Dutch for "cunt"), "*Godverdomme*," "*klote* "("bollocks"), "*Jezuz*," "*tering*" ("tuberculosis"), "*kanker*" ("cancer"), "*lul*" ("prick"), "*tyfus*" ("typhus"), and "bitch." But they did find that, even among a relatively homogeneous group of young female students, the use of swearing was extremely varied, with the reported number of swear words used per day ranging from zero to an impressively loquacious fifty. Not all women (and presumably not all men) are equally comfortable with—or motivated by—swearing.

While most research into swearing suggests that it is mainly used positively and constructively, the women who took part in this study said that they swear most often when they want to express negative emotions, then when they want to insult someone, and, in last place, when they want to express a positive emotion. Women who rated themselves as highly aggressive tended also to swear more but there was almost no connection with how satisfied they were with their lives.

Is there really a difference in the reasons why men and women swear? There might be. Men tend to feel comfortable using swearing in a jokey manner, as well as using it as a tool. For women, swearing is much more likely to be instrumental, to be used very carefully for effect. Professors Bailey and Timm found that the women they interviewed were also more likely to use swear words as a rhetorical device, to inject a bit of "punch" into their conversation. Said one interviewee: "I like 'that's fucked' or using 'fucking' as an adjective, and am impressed by others who can effectively

interject a curse into conversation. It makes for dynamism in communication."[20]

Dr. Karyn Stapleton says that women use swearing instrumentally to make an impression or ensure they are heard in mixed conversations—mainly because swearing is still seen as a "gender transgressive" act—it's still a hallmark of being one of the boys, especially if you're a girl.

"Among the women I've interviewed some of them definitely swear to subvert gender expectations: they're swearing to resist being seen as the good girl," she told me. However, that's not the whole story. Women also swear for much the same reasons that men do: to express or cover up their feelings, to make an impact, to raise a laugh, even as a form of politeness, but women run a much greater risk of social censure if a jocular bit of swearing goes flat.

"When you *ask* people about swearing, their first association is with aggression but their *use* of it is much more nuanced," says Dr. Stapleton. Women might say that swearing—even their own—is mainly negative but in practice we are capable of using it to be funny or inclusive: remember Ginette from the Power Rangers in chapter 4.

Ironically, despite its association with aggression, Karyn believes swearing might be a way of communicating pain and sadness for men and women alike: "That perceived association with verbal violence is one reason why swearing is more traditionally associated with a male speech style. However, swearing can actually be a way of covering up vulnerability for both genders."

I asked her if this might be another reason why the frequency of swearing is still higher for men than women in the various corpus studies that have been carried out. Men

are under greater social pressure to hide their vulnerability, whereas it's still more acceptable for women to talk about their feelings of uncertainty or hurt. There's definitely more research to be done but we do know that both women and men will adopt whatever kind of speech is expected from them in different social contexts. Both genders feel a pressure to conform, but in different ways.

"If you work in a context where swearing is the norm then women are going to swear, partly because they have to keep up with the men," Karyn told me. "But in interviews, women told me that they tend to use swearing for its social aspects—for humor and bonding. Women do a lot of this type of swearing." So rather than becoming more aggressive, we women are probably adopting more swearing as yet another way of being polite.

Women Are Bitches, Men Are Cunts: Gender-Specific Swearing

While women might be swearing alongside the men, men and women swear differently. In particular, the words we use about each other differ.

One of the big differences between men and women is the language we use about each other. In 2002 Claudia Berger from the University of Illinois looked at the ways in which male and female students insulted people, either to their faces or behind their backs.[21] She found that the students came up with more insulting words for men than for women (seventy-nine different male-specific swear words versus forty-six female-specific ones). Among those, the

bad language used about women and men is very different.

These students were more likely to use words that implied women were sexually incontinent ("slag," "whore"). Men, on the other hand, tended to attract swear words that suggest they aren't red-blooded heterosexuals who will shag any available woman. Men are most likely to be insulted with names that question their sexuality ("fag," "nancy") or masculinity ("cunt," "girl").

Dr. Susan Hughes's study of lower-working-class women in the north of England from the mid-1990s found that women rated "slag" and "slut" as far worse insults than "bitch," "cow" or even "cunt"—despite several decades of feminism, the surest way to insult a woman is still to imply that she is sleeping around.[22]

Why do we give women a hard time for having too much sex, and castigate men for having too little—or too little with women, at least? Professor Berger suggests that these swear words are so effective because they question the "honor and reputation" of the person on the receiving end. The different terms used to insult men and women reflect the age-old double standard about sex: male reputations are built on sexual prowess and women are still meant to be "good girls." It's power and purity raising its head again.

Swearing at women focuses on sexual behavior. The one exception—"bitch"—denigrates any woman who isn't "nice" enough. Swearing at men is much more varied. It includes insults based on parentage ("bastard"), intelligence ("shithead"), and parts of the body ("asshole," "cunt," "twat"). We also know that men are actually on the receiving end of more swearing than women.

Professor Tony McEnery has spent many years studying

the use of swearing in various databases of spoken and written language. He discovered that people are, in general, more likely to be abusive about their own sex, but while women were only slightly more likely to swear about a woman than they were about a man, men were significantly more likely to use swear words about men than women.* So, although women are responsible for nearly half the recorded swearing, they are on the receiving end of only just over a third of it.

My colleagues and I found a similar pattern, albeit about a different subject, when we studied swearing among soccer fans on Twitter. Fans are overwhelmingly more likely to swear about their own team and rarely slate the opposition.[23] In fact, the only player to be sworn about by the opposing team's fans was Kevin Nolan, then captain of West Ham, who was sent off and received a three-match ban after a particularly nasty foul on a Liverpool player. It's worth noting that Nolan was so roundly disliked at this point that even the West Ham fans couldn't conceal their delight, with tweets like "Nolan's last game for us? Let's fucking hope so! #WHUFC" and "FUCKING HAPPY DAYS KEVIN NOLAN IS GOING TO BE BANNED FOR 3 GAMES!!!!! GET IN THERE HAHAHA #thereisagod #whufc."

Women still tend to use the "weaker" swear words more often than men. As swear words lose power, we tend to see women using them more and men less. We don't know if

---

* Of 799 recorded episodes of women swearing, 407 were leveled at other women and 392 were aimed at men, while for men the difference is much greater: of 858 episodes of swearing, 702 were at or about other men and 156 at or about women.

this is because swear words adopted by women are seen as less powerful by men, and so used less by them, or because women become more willing to use these swear words after men have discarded them as weak. Professors Bailey and Timm did, however, discover that women will use "men's" swear words, but men rarely use "women's" swear words, which tend to be softer and more euphemistic. They might have been considered foul-mouthed in previous generations but it seems that as soon as a swear word falls out of the "strong" category, men cease using it. Of the twenty-nine uses of "darn," twenty-six were by women. Men didn't admit to using "Jeez," "shoot" or "crud" at all in the study. In their choice of swearing, as well as in the kind of swearing they attract, men's masculinity is a very vulnerable thing, it seems.

The idea that women are the purer sex when it comes to bad language seems to be a notion held by men about women, rather than a reflection of the truth. In reality, there is a lot of overlap between women's and men's speech depending on age, class, and context. As Susan Hughes points out, the (male) experts on the *Antiques Roadshow* tend to use a lot of "womanly" words like "pretty," "charming" or "delightful," while the women at Ordsall community center would be much more likely to use speech that might make a construction worker blush.

And it was a woman, after all, who paved the way for me to write this book without using a galaxy of asterisks. *Wuthering Heights* contains twenty-seven instances of "damn," "damned," or "damnable," twelve "God"s used as interjections, and insults like "slattenly witch" and "fahl, flaysom divil." In her preface to the book, Charlotte Brontë (writing under the pseudonym Currer Bell) first challenged the use of

"niceties" like "b——d" and "d—n" in novels. She explained that there would be no punches pulled in pursuit of realism:

> *Wuthering Heights* must appear a rude and strange production . . . A large class of readers . . . will suffer greatly from the introduction into the pages of this work of words printed with all their letters, which it has become the custom to represent by the initial and final letter only—a blank line filling the interval. I may as well say at once that, for this circumstance, it is out of my power to apologise; deeming it, myself, a rational plan to write words at full length. The practice of hinting by single letters those expletives with which profane and violent persons are wont to garnish their discourse, strikes me as a proceeding which, however well meant, is weak and futile. I cannot tell what good it does—what feeling it spares—what horror it conceals.

The negative attitudes we hold about women swearing are irrational and outdated and are, perhaps, finally starting to change. We should keep challenging those assumptions that insist that men must always be powerful and women must always be pure, and the way we use language has a profound impact on the way the sexes see each other.

Karyn Stapleton puts it most eloquently: "Swearing, for men and women, differs so much depending on the context. When women swear, they're expressing trust in other people. When you're able to relax with someone and swear with them, that means you trust them." Men do this too, but for women the message is particularly strong. "Swearing is still more circumscribed for women, it is a higher risk. But

that's changing slowly. The gender distinction is closing a bit, but not entirely. Not yet."

We should keep at it. Swearing is a powerful instrument, socially and emotionally. If women and men want to communicate as equals, we need to be equals in the ways in which we are allowed to express ourselves. Sod social censure. Let us allow men to cry and women to swear: we need both means of expression. I like this observation from British-American anthropologist Ashley Montagu, writing in the 1960s: "If women wept less they would swear more . . . Today instead of swooning or breaking into tears, she will often swear and then do whatever is indicated. It is, in our view, a great advance upon the old style."[24]

Too fucking right.

＊＊＊＊＊＊ 7

Schieße, Merde, Cachau:
Swearing in Other Languages

We learn early on in life that swearing is somehow special. Whether we notice an adult acting embarrassed after using a new word, or we get into trouble for saying particular things, we tend to learn sooner rather than later that some language is different. As children, the moment when we discover that some words have so much power is extremely exciting. Depending on the reaction of those around us we either internalize the taboos and resolve not to swear or we relish that power and resolve to use swear words as much as possible.

But in childhood we're still learning the social rules of language as well as its syntax and semantics. We learn the rules of use at the same time as we pick up the rules of grammar, without any formal instruction. What happens when we come to learn a second language? If we learn the *rules* of impolite behavior as small children, but learn the *language* of impolite behavior as an adolescent or an adult, can we ever truly understand swearing in a second language?

As anyone who has ever learned another language at secondary school will remember, the most well-thumbed pages of any bilingual dictionary are the ones with the dirty words. Many an unmotivated scholar, incapable of retaining the rudiments of the dative case, will commit to memory the German words "*Scheiß*" and "*Arschloch*" with ease. Adolescents are fascinated by taboos, and taboo words are no exception.

Adults who learn a second language might not make a beeline for the filthy words but exposure to popular culture while living abroad means that they will pick up taboo phrases whether they intend to or not. Studies of second-language learners show that adolescents are keen to learn bad language, young adults are keen to use it, and older adults pick it up—intentionally or otherwise—and can, with care, learn to use it appropriately.

But the languages we learn in infancy are the ones that will always have the greatest emotional resonance. New parents often find themselves reverting to the language of their own childhood when they bring home their firstborn; the tender endearments that they heard as infants sound more natural to use with their own babies than ones they learned later. We internalize the links between language and emotion most readily in childhood, which is why swearing in our first language tends to evoke stronger feelings than second-language swearing even for very fluent speakers.

How late can someone learn a second language and still have it mean the same—emotionally as well as intellectually—as their mother tongue? We know that people describe films, for example, in far more detached, factual ways when using their second language than they do when using their first.[1] Because of the emotional power of swearing, studying

its effect on second-language speakers is a great way to understand how and when the links between language and emotion are formed. These studies show that our personalities continue to evolve until late in life; languages we learn as adults, particularly ones that are learned while living in another country, have their own emotional resonance and can even unlock an entirely different personality.

Stepmother Tongues and the Strength of Language

Professor Jean-Marc Dewaele of Birkbeck, University of London, has made extensive studies of bad language and its emotional effects on the polyglot. He himself is intimidatingly multilingual, and he tells a story about the time he used a newly acquired swear word in his fourth language, Spanish.

"'When in Rome, do as the Romans' does not necessarily apply to swearing," he said. "I personally learned this lesson when using a taboo word in Spanish . . . after consumption of many *tapas* and red wine in one of Salamanca's bars. Although the exclamation '*joder*' ('fuck') had been uttered several times during the evening, my use of it was greeted by a stunned silence."[2]

That moment prompted Professor Dewaele's interest in swearing and emotion among people who speak multiple languages; a subject that he has spent over a decade researching. His findings show that how we learn a language, where we learn it, and even the way we were raised, can have profound effects on the way we swear. To try to get to the bottom of some of these effects, Dewaele asked over

1,000 multilinguals some questions about language and emotion. Using an online questionnaire he managed to find some remarkably gifted linguists: 144 bilinguals, 269 trilinguals, 289 quadrilinguals, and 337 pentalinguals—some of whom had been spoken to in two or three languages since birth.

When asked what language they choose to swear in, the respondents chose their first language significantly more often than their second, and their second significantly more often than their third. The large majority who preferred to swear in their first language did so for various reasons. For some it was the fact that first-language swearing is just more automatic. "K," a volunteer who speaks Finnish as a first language but also English, Swedish, and German, said, "If I would happen to hit myself with a hammer the words coming out of my mouth would definitely be in Finnish." Force of habit makes Finnish K's go-to language for spontaneous frustration swearing.

But it's not just habit that drives us to swear in our first language. Sometimes a little thought about the effect of our swearing can make us retreat to home turf. Many multilinguals say that they get a real feeling for the strength of their swearing only when using their native tongue. "Sandra"— who is a native German speaker but who has Italian as a second language—said, "If I am really angry only German words come into my mind; if I use Italian instead I may not use the right measure." The concern over using a swear word that is too strong (or too weak) can be a powerful motivation to stick with a language whose cultural rules we have internalized through experience early on in life.

As an adult, it's harder to pick up the cultural and

emotional resonance of swearing in a new language. That's because we learn how to interpret tone, gaze, facial expressions, and other emotional cues early in childhood and after a while this repertoire of emotional tells becomes somewhat fixed. In the 1980s, psychologist Professor Ellen Rintell recorded native English speakers expressing pleasure, anger, depression, anxiety, guilt, or disgust. She then asked speakers for whom English is a second language to identify the emotion in the recording, and to rate it for intensity. She found that even the most fluent speakers struggled to do as well on the task as people for whom English was their first language. That's not to say that we can't learn a new set of emotional signals, even as adults, but all of that instinctive emotional processing doesn't immediately translate to a second language.[3]

Learning a second language during or after adolescence tends to change the way in which we process feelings in that language. The words we learn in childhood have a profound ability to provoke or express emotion, whereas words we learn later on have a tendency always to feel more distant, less intense. Anglo-Canadian author Nancy Huston moved to France when she was at university. In her biographical essay *Nord perdu* (*Losing North*) she writes (in French):

In my case it is in French that I feel at ease in an intellectual conversation, in an interview, in a colloquium, in any linguistic situation that draws on concepts and categories learned in adulthood. On the other hand, if I want to be mad, let myself go, swear, sing, yell, be moved by the pure pleasure of speech, it is in English that I do it.

For Ms. Houston, her first language is the language that is more free, spontaneous, and emotionally expressive. She says that it's the language she chooses to swear in because the words come more readily to her. This is a common reaction, particularly when we let loose with interjection swearing: it's not just our familiarity with words that makes swearing so powerful, it's the fluency of our feelings.

But it's not just swearing that packs a greater emotional punch in your mother tongue. To test whether words learned in childhood set off stronger emotional reactions than words learned later in life, scientists at Boston and Istanbul Universities set out to measure the reactions of people who spoke both English and Turkish to the kind of telling-off they might have heard as children.[4] While these reprimands weren't in the form of swearing, they still have a lot in common with bad language: childhood reprimands tend to elicit shame and embarrassment on the part of the hearer; they are used to intensify an argument emotionally and, as we saw in chapter 1, childhood reprimands and swearing are the two categories of language that are most often retained, even after strokes that rob the sufferer of all other types of speech.

In their study, Professors Catherine Harris, Ayşe Ayçiçeği, and Jean Berko Gleason wired thirty-two native Turkish speakers to galvanic skin-response monitors. Importantly, none of these volunteers had learned English before the age of twelve, so all their tellings-off in childhood had been heard in Turkish. The scientists had them hear or read words that were neutral (e.g., "door"), positive (e.g., "joy"), negative (e.g., "disease"), taboo (e.g., "asshole"), and childhood scolds (e.g., "Don't do that!" and "Go to your room!").

The scientists found that the volunteers didn't react particularly strongly to the neutral, positive, or negative words, regardless of language. They reacted similarly strongly to the taboo words that they heard, regardless of whether they were in English or Turkish; their exposure to swearing in late adolescence had been enough to make English swearing an emotionally effective part of their language. However, the volunteers did respond very differently to the childhood reprimands depending on the language used. Even though the volunteers all understood the reprimands, their skin conductivity remained low—they showed no stress—when they heard the words in English. When they were exposed to the tellings-off in Turkish, and in particular when they heard rather than read them, their galvanic skin response went through the roof. Being told off in their native language was enough to make these volunteers (average age twenty-eight) break out in a cold sweat. This shows that understanding a word and feeling its emotional impact are two very different processes. We have to have experience of the emotional consequences of words if they are going to affect us.

It might still be possible to acquire a language fluently after childhood, but there is evidence that languages acquired after puberty never reach the same level of emotional force as languages learned earlier. Linguist Steven Kellman coined the phrase "stepmother tongue" for a language that has been acquired fluently but after early childhood.[5] What's more, the developmental stage we are at when we learn a language, particularly if we learn it immersively, might determine exactly which emotions we can access in it.

One of the first clues that emotions might be felt differently in a first and second language came from studies by the

psychologist Dr. Susan Ervin-Tripp in the 1950s and 1960s. It's a very human trait to tell ourselves stories, and part of what defines our personality are the types of stories we let ourselves believe. Dr. Ervin-Tripp wanted to see if people who spoke two languages fluently told different sorts of stories depending on the language they used, and the age at which they learned it. To do this she showed bilinguals various pictures and asked them to describe what was happening in them. The process was repeated six weeks apart, once in English and once in French.

When she showed the pictures to a group of sixty-four people who spoke French as a first language and had learned English as adults, the answers given in French were much more concerned with trying to be independent—a very adolescent preoccupation—whereas the descriptions they gave in English were more focused on achievement.[6] In a similar experiment with another group, when she showed a picture of a woman sitting on the floor with her head resting on the sofa to an American man who had been sent to a boarding school in Japan between the ages of eight and fourteen, she found that "in Japanese he suggested that it was a woman weeping over her lost fiancé and considering suicide, while in English he said that the picture depicted a girl finishing a project for a sewing class." This pattern was repeated over and over. When asked to describe a scene in Japanese, his answers were emotionally rich and dealt with themes such as loss and family. His answers in English, on the other hand, were much more impersonal. Japanese women living in San Francisco were also more likely to give very different answers when asked to complete sentences like "I will probably become . . ." When interviewed in English, their answers

were more likely to include careers ("a teacher" for example) whereas in Japanese they were more likely to say they would become mothers or housewives.[7]

Dr. Ervin-Tripp suggests that the psychological development we experience at the same time that we are learning a language becomes bound up in that language to some degree, and that part of our personality predominates as soon as we switch languages. For her volunteer, all the time that he was away from his family, and developing into puberty, he was communicating predominantly in Japanese. As he learned to process feelings of attachment and loss, he was also learning to express himself in Japanese, but these weren't feelings he had experienced to the same degree when he spoke English. For the rest of his life he could much more easily access these strong emotions in Japanese than English. The same is true for us all: as our emotional selves are developing, well into adolescence, we easily internalize emotional as well as literal meaning.

It's much harder for adults to internalize the emotional meaning of the words, but not impossible, especially in early adulthood as we're experiencing new challenges for the first time. Most of us are still learning how to respond to certain situations such as opening a bank account, renting a flat, or managing our finances well into early adulthood. As in childhood and adolescence, feelings experienced for the first time tend to become most closely linked with the language we are speaking at that stage in our lives. For example, "Johanna" is English but spent her young adulthood in Italy. "I'm more likely to express anger in Italian. Mainly because I've only really learned how to in the last few years and since I've spent my young adulthood here I've

gotten more practice raging at the government or the land-lord in my adopted language. I still end up feeling ridiculous when I get worked up about things in English."[8]

We know that this is generally the case thanks to an ingenious experiment by Professor Jeanette Altarriba from the State University of New York at Albany. She has spent many years studying the interaction between emotion and second languages using a variant of the Stroop test, the same one that was used to test Tourette's syndrome sufferers and their impulse control in chapter 3. In its original form the Stroop test tries to "trick" you into saying the written word rather than the ink color (page 78). The amount of time it takes to overcome the urge to give the correct-but-difficult answer rather than the wrong-but-automatic answer is a good measure of how hard you find it to inhibit unhelpful impulses. The test has been used to check for the effects of everything from aging to happiness to sleep deprivation.

But there's another way of using variants of the Stroop test. Rather than discovering how hard it is to overcome inhibition, we can use it to probe for incongruity. We know that people find incongruent stimuli much more challenging than congruent ones: it's easier to say "pink" if pink is both the color that is written and the ink that it is written in.

Professor Altarriba used a variant of the Stroop test as a probe to see how much emotion was packed into words in both a first and second language by testing for congruence between the emotion of the sentence and the words "posi-tive" and "negative."[9] She showed volunteers words on a screen and asked them to say "positive" if the word they read was a noun and "negative" if it was an adjective. The words

"positive" and "negative" were, in this case, acting like the color of the ink in the traditional Stroop test. However, some of those nouns and adjectives had positive emotional associations ("friend," "happy") while some were negative ("enemy," "angry"). Those associations act like the printed name of the color in the original Stroop test: the volunteers had to ignore the automatic associations in order to give the correct answer to the question "Is this a noun?"

Try it for yourself: how easy is it to answer "positive" for a noun and "negative" for an adjective with the following examples?

1. Joyful
2. Death
3. Fear
4. Beautiful
5. Sad
6. Treat
7. Fun
8. Frightening

The answers you should have given were 1) negative 2) positive 3) positive 4) negative 5) negative 6) positive 7) positive 8) negative. You probably found the second set of four words easier to classify than the first set of four. That's because we have to work much harder to say "positive" when we read a word like "death," which makes us feel decidedly negative.

Professor Altarriba carried out a detailed version of a similar experiment with some of her students. The thirty-two volunteers were bilinguals who spoke Spanish as a first

language, but who had been educated in English-speaking schools. Although Spanish was their mother tongue, these students were as close to truly bilingual as it is possible to be. Professor Altarriba found that, whether words were in English or Spanish, these volunteers found it easier to answer correctly when words were congruent (i.e., the noun "friend" demanding the response "positive" and the adjective "angry" demanding the response "negative") than when they were incongruent (i.e., the noun "death" demanding the response "positive" and the adjective "happy" demanding the response "negative").

The fact that the incongruent stimuli (emotionally negative nouns and emotionally positive adjectives) were much harder to process than the congruent ones means that the volunteers must have been feeling the emotional effect of the words strongly enough for it to interfere with the task of categorizing the words into nouns and adjectives. There would be no incongruity between how the word "death" makes you feel and the required answer ("positive") unless you'd internalized the emotional impact of that word.

But Professor Altarriba's students were unusual; very few of us are skilled enough to speak two languages with such fluency. Does the same effect hold for all second-language speakers? Professors Ayçiçeği and Harris, the ones who made graduate students tremble by playing childhood reprimands at them, somehow persuaded the same group of volunteers to try another language task, this time to do with memory.[10] Remember that these native Turkish speakers lived and worked in Boston but had not learned English until they got to the age of twelve. In the experiment they were shown lists of words that included taboo words in English

or Turkish like "asshole," "whore," "shit," and "*sevişmek*" ("fuck"), "*kahpe*" ("bitch"), "*fuhuş*" ("prostitution"). The list also included negative words such as "*katletmek*" ("murder"); positive words such as "*gülmek*" ("laugh"); neutral words such as "*masa*" ("table"); and those childhood reprimands again, such as "*Seni utanmaz!*" ("Shame on you!").

Professors Ayçiçeği and Harris told another of those little white lies that make psychology experiments possible. Before the experiment, they told the volunteers that they were being asked to rate the words from one to seven depending on how positive or negative they found them to be, but this isn't what they really wanted to test. After the "experiment" the true test began: they asked half of the volunteers to write down as many words as they could remember and asked the other half to circle the words that they had seen from a list of 128. None of the volunteers knew they would be quizzed beforehand and so hadn't been primed to make a special effort to remember the words.

The volunteers found it much harder to remember "bad" words in Turkish than they did in English. Even though these words were more familiar to them, being part of their mother tongue, the taboo words that made the volunteers feel bad were the words that were most easily forgotten. Taboo, negative, and scolding words were all more easily recalled when seen in English than when seen in Turkish, while there was no difference between the positive and neutral terms. Professor Ayçiçeği calls this an "emotional advantage" in the speakers' second language. Reprimands, taboos, and negative terms were far less hurtful in the second language and so were less likely to be deliberately forgotten.

That emotional advantage can stretch to finding greater self-expression in a second language, even if we are not as familiar with it. Sometimes we need to be able to express ourselves without being overwhelmed with feelings. "Some bilingual writers may prefer to write in a 'stepmother tongue,' escaping the emotional overcharge and traumatising powers of the mother tongue," says linguist Steven Kellman.[11] This is borne out by the experience of many of the multilinguals that Professor Dewaele has studied, who say that swearing is actually easier in their second language. While they might be more fluent in their first language, the emotional impact is far less in their second, particularly if they internalized very strong taboos while growing up.[12] For example, Nicole (a pseudonym) who speaks English as a first language, followed by German, French, Italian, and Spanish, told Professor Dewaele: "My parents were quite strict and I still have the phrase 'I'll wash your mouth out with soap and water' in my head! I'd never swear in English, but it's easier in German." Likewise, Maria, whose mother tongue is Spanish says, "I never swear in Spanish. I simply cannot. The words are too heavy and are truly a taboo for me."

Of course, sticking to your own language, especially if it's a rare one, can also give you the freedom to swear without getting into trouble. Didi, who speaks Sundanese, Bahsa Indo, and English, says that Sundanese has huge advantages as a language for swearing in. Not only does it feel strongest, as his mother tongue, it's also not very widely known, even in London. "I [swore at] somebody near Birkbeck College in 1997 using Sundanese while the person is English—it is . . . safer for me to do this."

How Universal Are Swear Words?

According to Professor Dewaele's studies, most of us do tend to stick to our first language when we swear. It not only feels more effective (and gives us a veneer of safety if we think we won't be understood) but also is the one we feel most emotionally competent in. "Language users seem to avoid use of linguistic nuclear devices if they are unsure about the yield," he says.[13]

Why can't we just translate the emotional intensity of a swear word in one language to another? Why are some swear words so much more offensive in some languages than others? To answer that, we need to look at how swear words have evolved in different languages.

Sometimes it's simply the case that one language has words that do a better job of expressing our emotions than another. Whether it's the wide range of Spanish swear words that had to be called in to replace a volley of ever-adaptable English "fucks" in the translation of *Pulp Fiction*, or something far more affectionate like the use of "*cariño*," which means tender, affectionate love in Spanish but has no direct English equivalent.

The experience of strong emotions—joy, pain, anger, fear—is shared by all humanity, regardless of our individual linguistic backgrounds. As a result, every culture needs strong words to express strong feelings. Likewise, every culture has taboos and these taboos make their way into the set of words that is considered "bad language." So far, so universal. But it isn't possible simply to substitute swear words from one language to another because the power of swearing is determined by culture: what is a mild reference in one language or culture might be the nuclear

option in another. As Sandra, who is bilingual in German and Italian, says in one of Professor Dewaele's studies: "Swearing in Italian means talking about God, Maria etc., in an obscene way, which in German doesn't mean a thing. The other way round in German you might use animals' names to insult a person. In Italian it wouldn't mean anything."

When learning a second language it can be very hard to pick up the best way of insulting or amusing someone. It's not simply a matter of having to be mindful of the emotional effects of language, it's often the case that we don't know the "right" words. What seems like a deadly insult at home might be totally ineffective in another country. Take this example from Professor Dewaele:

"Insults and [swearing] are highly culture-specific, as was highlighted again in March 2003 in the bitter verbal exchange between a Kuwaiti diplomat and an Iraqi minister where their respective mustaches became the target of insults. What is laughable in one culture might be deeply offensive in another."[14]

Even speakers of ostensibly the same language might use very different types of swearing.[15] For example, the difference between French that is spoken in France and French that is spoken in Canada is huge, especially when it comes to swearing and taboo words. French Canadians are much more likely to violate religious taboos when they want to swear. In French-speaking Canada, words like "hostie" ("host") and "vierge" ("virgin") are offensive: "hostie de voisin" is insulting in Quebec but meaningless in Lyon, where "salaud de voisin" ("bastard neighbor") would be a much more likely choice to convey the same sentiment.

Even in the same country, different regions, classes, and

social groups have different types of swearing. In chapter 6, on gender and swearing, we saw how differently working-class women in the north of England use swearing (plenty of sexual taboos but very little blasphemy, used to send strong social signals about their independence, page 159) compared to young women at college in the United States (lots of mild swearing, used to send strong social signals about acceptance, page 157). There has been no comprehensive census of swearing in every language, but we do know that they all include taboo words, and that those taboos vary both from place to place and over time.

"You taught me language; and my profit on't is, I know how to curse."

We tend not to be taught bad language in "formal" language lessons and so we end up acquiring it elsewhere: from books, films, the internet, friends. There are even speciality language guides, like *The Complete Merde!* by the pseudonymous "Geneviève," which translates swearing, cursing, and slang for the speaker of French as a second language, with phrases like "*Je m'en fous*" ("I don't give a fuck") and "*Il m'emmerde*" (he annoys me—literally "He gives me shit").

But the problem with swearing dictionaries is that they lack context. Swearing packs an emotional punch. The reason—we are told—that we aren't taught to swear in school is that there is always a "clean" alternative. But who seriously believes that "this is bad" and "this is shit" are remotely synonymous?

Given the huge differences in the cultural meaning of

swearing between different languages, surely it's more important to help users understand not only the literal but also the emotional and figurative meanings of swearing so that they don't make calamitous mistakes when trying to shift their swearing from one language to another?

Some language teachers, especially in secondary schools, are happy to turn the students' fascination with taboos to their advantage. An example from a study by Professor Monika Maria Chavez of the University of Wisconsin–Madison shows a teaching assistant (TA) encouraging his students to use insulting, joking, and informal language when talking to him.[16] His pupils take part in jocular abuse between themselves, and also jocular abuse of the TA:

FIRST STUDENT: *Du bist ein lustiger Mann. Deine schmutzige Kleidung war fantastisch. Kennst du Deine[sic] Tag mit keine [sic] Kreide an deine [sic] blaue [sic] Hose?* (You are a funny man. Your dirty clothes are wild. Do you know one single day without chalk on your blue pants [jeans]?)

SECOND STUDENT: *Dein grüner Hut ist super sexy.* (Your green hat is super sexy.)

THIRD STUDENT: *Du bist ein lustiger, verrückter TA, aber auch ein später TA.* (You are a funny, crazy TA but also a late TA.)

TEACHER: *Nur, nur wegen . . . den Scheißbussen na!* (Only, only because . . . of the shit buses!)

This TA says that he wants the students to get a feel for the real language, which is why he uses what he considers to be real terms. By insisting that students use German to say

the sorts of things that adolescent students say about their teachers—and by "rewarding" with some swearing in turn—he is encouraging them to use German as an instrument of self-expression, rather than teaching the language formally. But there is a difficulty here; many teachers feel uncomfortable with the idea of talking about the kinds of taboos that underpin swearing with their pupils, particularly on topics such as sex or race.

"These are difficult ethical questions. Just how much emotion-laden vocabulary and expressions should be taught to the learners? Should these words and expressions include the many synonyms referring to sexual anatomy and sexual behavior? What about words with racist connotations?" asks Professor Dewaele.[17]

According to linguist Robin-Eliece Mercury we risk doing students a great disservice unless we research ways of including swearing when we teach foreign languages.[18] Students are going to be exposed to swearing, cursing, and insults anyway through popular culture and through their friends, so surely it's better, she maintains, for them to learn how to swear responsibly? The problem is that there is very little research into how best to teach second-language learners about swearing. Ms. Mercury found herself unprepared when a student came to her to ask about swearing:

A female senior high school EFL student asked me about "bad words" in her weekly diary. She needed to understand what these words were used for, and why many American actors used them in movies. She inquired if it was acceptable for her to use them as well. My immediate knee-jerk reaction was to discourage

their use and to advise her not to discuss or think about taboo language . . . [H]ow uninformative that was, especially for a language learning student! So, via diary writing, we discussed what swear words were, and how problematic their use can be, even among native speakers. This student had legitimate questions about a part of English that exists but, unfortunately, is little spoken of in teaching contexts.

But, because swearing is such an important and powerful part of language, can we really say that we can speak a language until we can swear in it? While many people are careful about when and why they swear, anyone who lives in a country for any length of time is going to be exposed to that country's version of swearing, whether because they encounter anger and aggression or—more likely—because they want to be part of a group that shares an identity based on the kind of jocular abuse common among friends and colleagues. If we don't help students to learn the social rules of swearing, we're leaving them at a cultural disadvantage.

For me, it was the experience of films, football, and friends that taught me about swearing in French. Despite a first degree where I was not only taught French, but also spent a lot of time being taught in French, I didn't encounter the language in anything other than an educational or business context until I went out and began socializing with French people.

But maybe that's no bad thing. Swearing is so varied between peer groups, and is so culturally laden, that formal instruction might not be the best way for swearing to be learned. After all, in our first language we learn about swearing from our family and friends, and the cultural influences

that surround us. We know that our use of bad language isn't determined solely by the language we are speaking and the country we are in. We tend to learn habits of swearing that are distinct to the social groups we belong to and the identities that we feel we have. If we receive any formal instruction about swearing at all it is usually limited to the idea that swearing is offensive and shouldn't be used. We have to learn about swearing in the context of genuine culture.

For example, children learning a second language from their peers are usually fairly adept at picking up "dirty" or "bad" words—perhaps because their peers have such dreadful poker faces. Professor Ben Rampton of King's College London studied children picking up Punjabi in the playground from their friends.[19] He equipped children with radio microphones to study their speech. In one example David, who speaks English as a first language, is sitting with Jagdish, one of his best friends, and some other boys, all of whom speak Punjabi. David is twelve and Mohan and Jagdish are thirteen:

> Jagdish asks David (in Punjabi), "Do you want to bum Laura?" and the Punjabi-speaking boys laugh.
>
> David replies, "No, I don't think so."
>
> Jagdish says, "No, I said . . . that means, 'Are you going to beat Laura up?'"
>
> David: "No."
>
> Sukhbir: "Yes it does."
>
> David: "No, it means, 'Are you going to make her pregnant?'"

David doesn't exactly know what Jagdish's Punjabi colloquialism means—from my understanding, pregnancy

doesn't result from "bumming" someone. It's possible that even Jagdish doesn't *really* know what he's talking about. But David is smart enough to guess from the gales of laughter from the other boys that he is being tricked and, being on the cusp of adolescence, suspects that the trick is something to do with girls and sex. He won't believe Jagdish and Sukhbir's translation and tries to use context and his imagination to guess what was said.

This is a very natural way of learning language, and especially of learning how to swear. Because David learned this phrase in a context where there were overtones of "dirtiness" and "taboo," and where there was a strong emotional context of piss-taking and male bonding, it's likely that he'll have a far better understanding of what the phrase really means, and where it is and is not appropriate to use it, than if he had simply learned it from a book.

Pulp Friction: The Problems of Translation

One of the best ways to learn a language is to immerse yourself in its culture. This means reading the books and websites, watching the films, television, and videos, and listening to the songs that are popular in the culture you're trying to be a part of. We acquire those little "tells" about the social group we belong to by assimilating what others do, by mimicking the kinds of swearing, jokes, slang, and references that they use.

But what happens when we need to move something from one culture to another? When those very same cultural

influences that teach us how to belong to a specific group are being repackaged for another audience? That's where the work of a translator comes in. While some words might translate directly: for example "shit" into "*merde*," "*sheiße*," or "*mierda*," there is no guarantee that the words "mean" the same thing: the degree of offensiveness, the frequency of use, and the type of situations where the word is acceptable all differ from place to place.

The job of a translator is not simply to substitute one word for another based on what the bilingual dictionary says. This is one reason why machine translation is so challenging. Good translations preserve the *sense* of what's being said, emotionally and culturally as well as semantically. And swearing presents a particular challenge to translators, partly because it's such an emotive part of language, so preserving that emotional force really matters, but also because different cultures have such different taboos, and have very different ways of using swearing.

Will McMorran and Thomas Wynn produced a new translation of the Marquis de Sade's *12 Days of Sodom* for Penguin Classics in 2016. "We had a duty to be just as rude, crude, and revolting as Sade," said McMorran.[20] "Was a *vit* a prick, dick or cock? Were *tétons* boobs, tits or breasts? Was a *derrière* a behind, backside, or indeed a *derrière*?" In the end they chose to use current sexual slang terms, as long as they didn't sound jarringly contemporary.

But there is more to translating profanity than just choosing *le mot juste*. Cross-cultural studies of English and Spanish swearing show that English speakers tend to stick to a few well-established swearing phrases whereas Spanish speakers tend to be far more inventive in the swearing that they use.

For example, the translators of *Pulp Fiction* into Spanish had to work hard to translate all of the instances of "fuck" in the script. *Pulp Fiction*'s script clocks in at 1.74 fucks-per-minute. For context, *Scarface* manages a mere 1.21 fpm, while *Hot Tub Time Machine* racks up 2.12 fpm. About one-sixtieth of the running time of *Pulp Fiction* is made up of swearing and the majority of that swearing uses "fuck" in some sense. That's a significant chunk of the film that's made up of f-words and yet it doesn't get monotonous. That's because, in English, "fuck" can be used as a verb ("Fuck you"), adjective ("It's fucked"), noun ("I don't give a fuck"), and either literally ("We fucked"), figuratively ("Don't fuck with me"), or as an interjection on its own ("FUCK!")

In Spanish, however, there is no single word that can form such feats of lexical gymnastics. The translator had to inject a much wider variety of vocabulary into the script. Dr. Ana Maria Fernández Dobao from the University of Montreal made an in-depth analysis of the numerous ways in which the translators had to come up with a whole range of swearing when they prepared the film for the Spanish-speaking market.[21]

Dr. Fernández Dobao explains that, in Spanish, the literal and figurative uses of "fuck" are somewhat distinct. "*Follar*" is a vulgar word that literally relates to sex, whereas *joder* can be used in some places where "fuck" is used metaphorically. For example, in English, the following two lines of dialogue use "fuck her/him":

1. "Fuck him! Scotty, he's a better boxer he still be alive."
2. "What did he do? Fuck her?"

The first one, is translated using *joder*:

1. "*¡Que se joda! Scotty, si hubiese sido buen boxeador aún viviría.*"

And the second, where the inquiry is literally "Did he have sex with her?" uses *follar*:

2. "*¿Qué hizo? ¿Follársela?*"

However, since there's no form of *joder* that will work as an adjective, this means that sentences like "Yes! I've fuckin' looked!" have to be switched around to make "*¡Sí joder! ¡He mirado!*" ("Yes, fuck it! I've looked!")

The use of "fuck" as a metaphorical verb ("fuck this, fuck that, fuck you") doesn't translate at all with *joder*, however. Spanish has a wider variety of ways of expressing this frustration. For example:

"Have you ever given a guy a foot massage?"
"Fuck you!"
"How many?"
"Fuck you!"

Becomes:

"*¿Te importaría masajearle los pies a un hombre?*"
"*¡Vete al cuerno!*" ("Go to the horned one"—sort of like "Go to hell!")
"*¿Has hecho muchos?*"
"*¡Vete al cuerno!*"

While

"Fuck pride! Pride only hurts, it never helps you, fight through that shit.'

becomes

"¡A la mierda el orgullo! [Go to shit with pride] El orgullo sólo hace daño, no te ayudará jamás, lucha contra esa mierda!"

And the interjection "Where the fuck is it?" becomes "¿Dónde cojones está?" or "Where the bollocks is it?"

It's not just "fuck" that doesn't translate directly. Motherfucker isn't used as an insult. Instead, the idea that a man's wife is fucking around behind his back is a much more common slight with cabrón or cabronazo (cuckold) used instead. "English, motherfucker, do you speak it?" becomes "Mi idioma cabronazo ¿sabes hablarlo?"

Dr. Fernández Dobao says English is much richer in the many different ways that it uses the same few swear words but that Spanish is more creative. Simply translating *Pulp Fiction* so that it has 1.74 *joders* per minute would have sounded monotonous and nonsensical. Instead, the translators had to press a much wider range of swear words into service in order to keep up the same level of emotional intensity.

Some translators feel the need to shy away from using so much swearing in order to make their audience feel comfortable with the translated text. Markus Karjalainen from the University of Helsinki studied two Swedish translations of *The Catcher in the Rye*—one from 1953 and one from 1987. Of the 778 swear words in the original text, the translators

omitted 469 in 1953 and 471 in 1987. Both translators left out a lot of the uses of "damn," "goddam," and "hell" that make up most of the original book's swearing, and they also dropped a lot of the next most common swear word, "bastard." For example, "I mean I'm not going to be a goddam surgeon or a violinist or anything anyway" becomes "*Jag tänker i alla fall inte bli nån gammal kirurg eller violinist eller så*" (1953) and "*Jag tänker inte bli vare sig kirurg eller violinist så det spelar ingen roll*" (1987), both of which roughly translate to "I don't think I'll end up as [some old] surgeon . . ."

Even though there are perfectly serviceable swear words for all of the omissions in the two translations of *Catcher in the Rye*, the translators prefer to keep the swearing to a minimum. According to Mr. Karjalainen, the Swedish temperament means that a little swearing goes a lot farther than it does in the English of J. D. Salinger.

"When I think of Sweden and its people, the first word that comes to mind is 'lagom.'* [For most average Swedes] moderation is a virtue and extremes in either direction should be avoided," says Mr. Karjalainen. "Lagom can be a positive thing, but it can also lead to an apparent superficiality on many levels. Not saying what one really means; suppressing emotional language to great lengths to avoid contrast and conflict."

The meaning of language goes beyond the dictionary definitions of words used: the spirit and sense of how we swear changes as we switch from culture to culture and it's easy for this sense to be lost in translation, even when it's the

---

* While there's no direct translation into English, *lagom* carries the sense of "enough" or "just right." It is a word that reflects a national pride and preoccupation with moderation.

same person writing. Julien Green, a bilingual author, was partway through writing a book in his native French when his publisher asked if he would, instead, write it in English. He intended to simply translate what he had already written from French to English. However:

> Rereading what I had written [I] realized that I was writing another book, a book so different in tone from the French that the whole aspect of the subject must, of necessity, be altered. It was as if, in writing English, I had become another person . . . There was so little resemblance between what I wrote in English and what I had already written in French that it might almost be doubted that the same person was the author of these two pieces of work.[22]

As we saw in chapter 6, even within a single country the differences in swearing can be significant to the point where you can guess the gender of a speaker just from the type of swearing that they use (page 156). However, the differences between languages are more significant. Different cultures have different taboos, whether it's the religious swearing of Italian or Canadian French, or the mustache-based insults of the Middle East. This poses a huge challenge for language learners and translators alike.

When we move between French, Japanese, and English, the words "*chat*," "*neko*," and "cat" all signify more or less the same thing. But the way we express ourselves emotionally, even the types of emotion that it is permissible to express, vary widely. Language is culturally laden, and no part of language more so than swearing. The effect of swearing is dependent on the emotional states we experience when

we see, hear, or use them and this emotional effect can be either freeing or overwhelming, depending on the kind of taboos that we have internalized.

And we can't simply look up the dirty words in a dictionary: not that that ever stops me. The only way to truly understand the force of swearing in another language is to experience it, in real life or from real culture. We have to immerse ourselves in feelings in order to learn the true meaning of swearing, and perhaps that's why some of us find it harder to pick up the nuances of language as we get older.

"To become skilled, one needs a lot of practice,' says Dr. Dewaele.[23] "After the sociopragmatic *faux pas* with my Spanish friends [over his use of *joder*], I decided to limit my swearing in Spanish to Captain Haddock's favorite expression 'rayos y truenos.'"* Even the most fluent of linguistics professors finds it hard to translate his feelings into the right type of bad words.

---

* "Thunder and lightning," here used as a mild outburst, in the same way that the English version of Captain Haddock says, "Blistering barnacles!"

Conclusion

Imagine our very earliest ancestors as they first experimented with the basics of language. This new cognitive tool opened up a whole new world of collaboration: language allows you to transfer knowledge, plans, and desires from your own mind to someone else's. Our ancestors can make effective plans for collaborative hunting ("You go and scare that antelope so it runs over here and we'll catch it as it goes past"). They learned how to pass on vital skills to their children ("Never grab a hissing vine"), they even learned to think out loud ("Do you think the antelope might be fatter in the next valley? And fewer snakes?").

Something else happens at the same time. Imagine those ancestors planning their antelope hunt. "I'm tired of antelope!" says one. "Can't we have hippo?" "Hippos are really dangerous and hard to catch!" "But antelopes are stringy and there's barely enough meat for a week!" Tempers start to flare.

In one cave, the curseless cave, either the hunters go

their separate ways or they resort to violence. In the first instance, one fails to catch the speedy antelope and goes hungry, the other is trodden on by an angry hippopotamus, gets gangrene, and dies. In the second, the disagreement becomes a physical fight, at the end of which neither is in any fit state to catch anything.

In the second cave, they're trying something new. They have a powerful set of words that refer to things and behaviors that cause shame or fear. Maybe it's to do with illness ("May your leg go septic after being trodden on by a hippo") or bodily functions ("You are as welcome in this cave as antelope turds") or the deities that cause the sun to rise or the crops to ripen. These words have real emotive force. Instead of coming to blows, the dwellers in this cave fling insults and curses. No bones are broken and everyone is fit for the hunt at the end of it.

Evolutionary psychology is a bit like the *Just So* stories: there's no way to know for sure that any of this happened. Using taboo words might have caused more fights, led to people being ostracized, splintered tribes. Our ancestors might not have developed swearing anywhere near as precociously as our chimpanzee cousins. But the idea of swearing as a social tool certainly hangs together. It's consistent with what we know of our history; that banding together in bigger societies is both stressful and necessary to develop the culture that we have. It's consistent with what we observe today; swearing helps us bear pain, work together, and communicate emotions.

I started this book by saying that I didn't want to encourage you to swear more. Swearing is like mustard; a great ingredient but a lousy meal. We need that part of our

language to keep its potency, its slightly risky nature, otherwise it wouldn't be swearing. We drop those words that don't give us a sufficiently strong punch anymore and pick up words that have taken on the mantle of the unsayable.

We will never know for sure where swearing came from, but we do know that we keep reinventing it, just when it seems to be losing its power. We need swearing and, however we might have invented it, I'm fucking glad that we did.

Acknowledgments

My thanks first go to Team Science Baby, David, Mike, Romilly, and Rosetta, whose patience, support, love, and motivation helped me all the way.

If I have sworn more, it is because I stand on the shoulders of giants. Thank you to all of the researchers whose work has intrigued, amazed, and moved me while I've been writing this book. Any errors of interpretation are entirely my responsibility. Thank you in particular to Richard Stephens, Barbara Plester, Janet Holmes, Karyn Stapleton, Megan Robbins, James Saunders, and James O'Connor for their fascinating discussions about their research and other work.

Thank you to Gordon Wise and Peter Tallack, who offered early thoughts and critique on the initial proposal. Immense gratitude to Carrie Plitt, for keeping her eye on the left field and for guiding me so diligently through polishing and pitching. Likewise, to Rebecca Gray for the notes that shaped not just the words in the book, but the way that I approach writing. You are the best teachers I could wish for. Thank you to Trevor Horwood who polished, pointed, and perfected throughout.

Thank you to George Lucas for a very enjoyable walking tour of New York, and to Matt Weiland for graciously letting me get my geek on about the WPA.

Thank you to Rhoda Baxter and Dan Smith for sharing their tales from the trenches. And finally to Fred Johnson: "Yes, the book is finished." Thank you for asking.

Notes

Introduction: What the Fuck Is Swearing?

1. G. Hughes, *Swearing: A Social History of Foul Language, Oaths and Profanity in English*. Blackwell, 1991.
2. E. Byrne and D. Corney, Sweet FA: Sentiment, Soccer and Swearing. In S. P. Papadopoulos et al. (eds.): *Proceedings of the SoMuS ICMR 2014 Workshop, Glasgow, Scotland*, 01-04-2014, published at http://ceur-ws.org.
3. B. K. Bergen, *What the F: What Swearing Reveals about Our Language, Our Brains, and Ourselves*. Basic Books, 2016.
4. E. Welhoffer, Strafe Für Beleidigungen: Wie Teuer Ist Der "Stinkefinger"? *Express.de*, 15 March 2016. www.express.de/news/politik-und-wirtschaft/recht/beleidigung-beschimpft-strafe-schimpfwort-teuer-anzeige-strafbar-1261268-seite2.
5. Rechtspraak.nl (database of Dutch court cases and rulings): https://uitspraken.rechtspraak.nl/#zoekverfijn/ljn=BD2881.
6. M. Mohr, *Holy Sh*t: A Brief History of Swearing*. Oxford University Press, 2013.
7. Hughes, *Swearing*.
8. Ofcom. Attitudes to Potentially Offensive Language and Gestures on TV and Radio. www.ofcom.org.uk/research-and-data/tv-radio-and-on-demand/tv-research/offensive-language-2016.

9. S. Pinker, What the F***? *New Republic*, October 8, 2007. https://newrepublic.com/article/63921/what-the-f.

10. K. Sylwester and M. Purver, Twitter Language Use Reflects Psychological Differences between Democrats and Republicans. *PLoS ONE* 10, 2015, e0137422. doi:10.1371/journal.pone.0137422.

11. K. L. Jay and T. B. Jay, Taboo Word Fluency and Knowledge of Slurs and General Pejoratives: Deconstructing the Poverty-of-Vocabulary Myth. *Language Sciences* 52 (2015), 251–259. doi:10.1016/j.langsci.2014.12.003.

12. G. Rayner, Sir Winston Churchill Quotes: The Famous Lines That He Never Said. *Telegraph*, October 13, 2014. www.telegraph.co.uk/news/politics/conservative/11155416/Sir-Winston-Churchill-the-famous-lines-that-he-never-said.html.

Chapter 1: The Bad Language Brain: Neuroscience and Swearing

1. P. Ratiu and I.-F. Talos, The Tale of Phineas Gage, Digitally Remastered. *New England Journal of Medicine* 351 (2004), e21. doi:10.1056/NEJMicm031024.

2. D. Van Lancker and J. Cummings, Expletives: Neurolinguistic and Neurobehavioral Perspectives on Swearing. *Brain Research Reviews* 31 (1999), 83–104. doi:10.1016/S0165-0173(99)00060-0.

3. D. Van Lancker and K. Klein, Preserved Recognition of Familiar Personal Names in Global Aphasia. *Brain and Language* 39 (1990), 511–529. doi:10.1016/0093-934X(90)90159-E.

4. L. J. Speedie et al., Disruption of Automatic Speech following a Right Basal Ganglia Lesion. *Neurology* 43 (1993), 1768–1768. doi:10.1212/WNL.43.9.1768.

5. R. L. Heath and L. X. Blonder, Spontaneous Humor among Right Hemisphere Stroke Survivors. *Brain and Language* 93 (2005), 267–276. doi:10.1016/j.bandl.2004.10.006.

6. P. Shammi and D. T. Stuss, Humour Appreciation: A Role of the Right Frontal Lobe. *Brain* 122 (1999), 657–666. doi:10.1093/brain/122.4.657.

7. G. Gainotti, Unconscious Processing of Emotions and the Right Hemisphere. *Neuropsychologia* 50 (2012), 205–218. doi:10.1016/j.neuropsychologia.2011.12.005.

8. A. Öhman et al., Emotion Drives Attention: Detecting the Snake in the Grass. *Journal of Experimental Psychology: General* 130 (2001), 466–478. doi:10.1037/AXJ96-3445.130.3.466.

9. T. Indersmitten and R. C. Gur, Emotion Processing in Chimeric Faces: Hemispheric Asymmetries in Expression and Recognition of Emotions. *Journal of Neuroscience* 23 (2003), 3820–3825.

10. E. Hitchcock and V. Cairns, Amygdalotomy. *Postgraduate Medical Journal* 49 (1973), 894–904. doi:10.1136/pgmj.49.578.894.

Chapter 2: "Fuck! That hurts." Pain and Swearing

1. R. Stephens, J. Atkins, and A. Kingston, Swearing as a Response to Pain. *Neuroreport* 20 (2009), 1056–1060. doi:10.1097/WNR.0b013e32832e64b1.

2. P. M. Aslaksen et al., The Effect of Experimenter Gender on Autonomic and Subjective Responses to Pain Stimuli. *Pain* 129 (2007), 260–268. doi:10.1016/j.pain.2006.10.011.

3. J. L. Rhudy and M. W. Meagher, Negative Affect: Effects on an Evaluative Measure of Human Pain. *Pain* 104 (2003), 617–626. doi:10.1016/S0304-3959(03)00119-2.

4. E. Kandel, J. Schwartz, and T. Jessell, *Principles of Neural Science.* McGraw-Hill Medical, 2000.

5. N. K. Lowe, The Nature of Labor Pain. *American Journal of Obstetrics and Gynecology* 186 (2002), S16–S24. doi:10.1016/S0002-9378(02)70179-8.

6. T. Saisto et al., Reduced Pain Tolerance during and after Pregnancy in Women Suffering from Fear of Labor. *Pain* 93 (2001), 123–127. doi:10.1016/S0304-3959(01)00302-5.

7. A. E. Williams and J. L. Rhudy, Emotional Modulation of Autonomic Responses to Painful Trigeminal Stimulation. *International Journal of Psychophysiology* 71 (2009), 242–247. doi:10.1016/j.ijpsycho.2008.10.004.

8. R. Stephens and C. Allsop, Effect of Manipulated State Aggression on Pain Tolerance. *Psychological Reports* 111 (2012), 311–321. doi:10.2466/16.02.20.

9. R. Stephens and C. Umland, Swearing as a Response to Pain— Effect of Daily Swearing Frequency. *Journal of Pain* 12 (2011), 1274–1281. doi:10.1016/j.jpain.2011.09.004.

10. DeWall N. C. et al., Acetaminophen Reduces Social Pain: Behavioral and Neural Evidence. *Psychological Science* 21 (2010), 931–937. doi:10.1177/0956797610374741.

11. T. Deckman et al., Can Marijuana Reduce Social Pain? *Social Psychological and Personality Science* 13 (2013), 60–68. doi:10.1177/1948550613488949.

12. Kandel, Schwartz, and Jessell, *Principles of Neural Science*.

13. M. J. Bernstein and H. M. Claypool, Social Exclusion and Pain Sensitivity: Why Exclusion Sometimes Hurts and Sometimes Numbs. *Personality and Social Psychology Bulletin* 38 (2012), 185–196. doi:10.1177/0146167211422449.

14. L. Lombardo, Hurt Feelings and Four Letter Words: The Effects of Verbal Swearing on Social Pain. Honors Thesis, School of Psychology, The University of Queensland, 2012.

15. S. Seymour-Smith, "Blokes Don't Like That Sort of Thing": Men's Negotiation of a "Troubled" Self-help Group Identity. *Journal of Health Psychology* 13 (2008), 785–797. doi:10.1177/1359105308093862.

16. S. Seymour-Smith, Illness as an Occasion for Storytelling: Social Influences in Narrating the Masculine Self to an Unseen Audience. In C. Horrocks, K. Milnes, and G. Roberts (eds), *Narrative, Memory and Life Transitions*. Huddersfield University Press, 2002.

17. M. L. Robbins et al., Naturalistically Observed Swearing, Emotional Support, and Depressive Symptoms in Women Coping with Illness. *Health Psychology* 30 (2011), 789–792. doi:10.1037/a0023431.

Chapter 3: Tourette's Syndrome, or Why This Chapter Shouldn't Be in This Book

1. T. Jay, *Why We Curse: A Neuro-Psycho-Social Theory of Speech*. John Benjamins, 2000.

2. www.cdc.gov/ncbddd/tourette/diagnosis.html.

3. M. H. Bloch and J. F. Leckman, Clinical Course of Tourette Syndrome. *Journal of Psychosomatic Research* 67 (2009), 497–501. doi:10.1016/j.jpsychores.2009.09.002.

4. S. Wilhelm et al., Randomized Trial of Behavior Therapy for Adults with Tourette Syndrome. *Archives of General Psychiatry* 69 (2012), 795–803. doi:10.1001/archgenpsychiatry.2011.1528.

5. C. A. Conelea, S. A. Franklin, and D. W. Woods, Tic, Tourettes, and Related Disorders. In R. J. R. Levesque (ed.), *Encyclopedia of Adolescence*. Springer New York, 2011, pp. 2976–2983.

6. Jay, *Why We Curse*, p. 65.

7. A. E. Lang, E. Consky, and P. Sandor, "Signing Tics"—Insights into the Pathophysiology of Symptoms in Tourette's Syndrome. *Annals of Neurology* 33 (1993), 212–215. doi:10.1002/ana.410330212.

8. R. M. Kurlan, Treatment of Tourette Syndrome. *Neurotherapeutics* 11 (2013), 161–165. doi:10.1007/s13311-013-0215-4.

9. K. J. Black et al., Progress in Research on Tourette Syndrome. *Journal of Obsessive-Compulsive and Related Disorders* 3 (2014), 359–362. doi:10.1016/j.jocrd.2014.03.005.

10. J. Piacentini et al., Behavior Therapy for Children with Tourette Disorder: A Randomized Controlled Trial. *Journal of the American Medical Association* 303 (2010), 1929–1937. doi:10.1001/jama.2010.607.

11. Conelea et al., Tic, Tourettes, and Related Disorders.

12. R. Wadman, V. Tischler, and G. M. Jackson, "Everybody just thinks I'm weird": A Qualitative Exploration of the Psychosocial Experiences of Adolescents with Tourette Syndrome. *Child Care Health and Development* 39 (2013), 880–886. doi:10.1111/cch.12033.

13. R. F. Baumeister, K. D. Vohs, and D. M. Tice, The Strength Model of Self-Control. *Current Directions in Psychological Science* 16 (2007), 351–355. doi:10.1111/j.1467-8721.2007.00534.x.

14. R. Elliott, Executive Functions and Their Disorders: Imaging in Clinical Neuroscience. *British Medical Bulletin* 65 (2003), 49–59. doi:10.1093/bmb/65.1.49.

15. J. R. Stroop, Studies of Interference in Serial Verbal Reactions. *Journal of Experimental Psychology* 18 (1935), 643–662. doi:10.1037/h0054651.

16. S. Palminteri et al., Dopamine-Dependent Reinforcement of Motor Skill Learning: Evidence from Gilles de la Tourette Syndrome. *Brain* 134 (2011), 2287–2301. doi:10.1093/brain/awr147.

17. Kurlan, Treatment of Tourette Syndrome.

18. R. P. Michael, Treatment of a Case of Compulsive Swearing. *British Medical Journal* 1 (1957), 1506–1508.

19. D. Van Lancker and J. Cummings, Expletives: Neurolinguistic and Neurobehavioral Perspectives on Swearing. *Brain Research Reviews* 31 (1999), 83–104. doi:10.1016/S0165-0173(99)00060-0.

20. A. Clempson, S. Dobson, and Judith S. Stern, P7 Dentists Treating Tourette Syndrome. *Journal of Neurology Neurosurgery and Psychiatry* 83 (2012). doi:10.1136/jnnp-2012-303538.24.

21. R. J. Maciunas et al., Prospective Randomized Double-Blind Trial of Bilateral Thalamic Deep Brain Stimulation in Adults with Tourette Syndrome. *Journal of Neurosurgery* 107 (2007), 1004–1014. doi:10.3171/JNS-07/11/1004; M. S. Okun, K. D. Foote, and S. S. Wu, A Trial of Scheduled Deep Brain Stimulation for Tourette Syndrome: Moving Away from Continuous Deep Brain Stimulation Paradigms. *Archives of Neurology* 70 (2013), 85–94. doi:10.1001/jamaneurol.2013.580.

22. L. Ackermans et al., Double-Blind Clinical Trial of Thalamic Stimulation in Patients with Tourette Syndrome. *Brain* 134 (2011), 832–844. doi:10.1093/brain/awq380.

23. A. Duits et al., Unfavourable Outcome of Deep Brain Stimulation in a Tourette Patient with Severe Comorbidity. *European Child and Adolescent Psychiatry* 21 (2012), 529–531. doi:10.1007/s00787-012-0285-6.

24. Kurlan, Treatment of Tourette Syndrome.

25. Wilhelm et al., Randomized Trial of Behavior Therapy for Adults with Tourette Syndrome.

26. Peterson was discussing Black et al., Progress in Research on Tourette Syndrome.

27. Piacentini et al., Behavior Therapy for Children with Tourette Disorder.

28. Wilhelm et al., Randomized Trial of Behavior Therapy for Adults with Tourette Syndrome.

29. H. Smith et al., Investigating Young People's Experiences of Successful or Helpful Psychological Interventions for Tic Disorders: An Interpretative Phenomenological Analysis Study. *Journal of Health Psychology* 21 (2016), 1787–1798. doi:10.1177/1359105314566647.

Chapter 4: Disciplinary Offense: Swearing in the Workplace

1. B. Plester and J. Sayer, "Taking the Piss": Functions of Banter in the IT Industry. *Humor* 20 (2007), 157–187. doi:10.1515/HUMOR.2007.008.

2. J. V. O'Connor, *Cuss Control: The Complete Book on How to Curb Your Cursing.* Three Rivers Press, 2000.

3. N. Daly et al., Expletives as Solidarity Signals in FTAs on the Factory Floor. *Journal of Pragmatics* 36 (2004), 945–964. doi:10.1016/j.pragma.2003.12.004.

4. M. Haugh and D. Bousfield, Mock Impoliteness, Jocular Mockery and Jocular Abuse in Australian and British English. *Journal of Pragmatics* 44 (2012), 1099–1114. doi:10.1016/j.pragma.2012.02.003.

5. Kate Fox, *Watching the English: The Hidden Rules of English Behaviour.* Hodder & Stoughton, 2004.

6. C. Scherer and B. Sagarin, Indecent Influence: The Positive Effects of Obscenity on Persuasion. *Social Influence* 1 (2006), 138–146. doi:10.1080/15534510600747597.

7. G. Feldman et al., We Do Give a Damn: The Relationship between Profanity and Honesty. *Social Psychological and Personality Science* (published online January 2017). doi:10.1177/1948550616681055.

8. M. L. Newman et al., Lying Words: Predicting Deception from Linguistic Styles. *Personality and Social Psychology Bulletin* 29 (2003), 665–675. doi:10.1177/0146167203029005010.

9. E. Rassin and S. van der Heijden, Appearing Credible? Swearing Helps! *Psychology, Crime and Law* 11 (2005), 177–182. doi:10.1080/1068316 05160512331329952.

10. K. L. Jay and T. B. Jay, Taboo Word Fluency and Knowledge of Slurs and General Pejoratives: Deconstructing the Poverty-of-Vocabulary Myth. *Language Sciences* 52 (2015), 251–259. doi:10.1016/j.langsci.2014.12.003.

Chapter 5: "You damn dirty ape." (Other) Primates that Swear

1. G. J. Romanes, *Mental Evolution in Man.* D. Appleton and Co., 1889.

2. R. M. Yerkes, *Almost Human.* Jonathan Cape, 1925.

3. W. Kellogg and L. Kellogg, *The Ape and the Child*. Hafner Publishing, 1933.

4. H. S. Terrace, *Nim: A Chimpanzee Who Learned Sign Language*. Columbia University Press, 1986, p. 137.

5. M. R. Lepper, D. Greene, and R. E. Nisbett, Undermining Children's Intrinsic Interest with Extrinsic Reward: A Test of the "Overjustification" Hypothesis. *Journal of Personality and Social Psychology* 28 (1973), 129–137. doi:10.1037/h0035519.

6. D. Morris, *The Biology of Art: A Study of the Picture-Making Behaviour of the Great Apes and Its Relationship to Human Art*. Methuen, 1962, pp. 158–159.

7. R. A. Gardner and B. Gardner, *The Structure of Learning: From Sign Stimuli to Sign Language*. Lawrence Erlbaum, 1998.

8. Ibid., p. 292.

9. R. Fouts and S. Mills, *Next of Kin: What My Conversations with Chimpanzees Have Taught Me about Intelligence, Compassion and Being Human*. Michael Joseph, 1997, p. 25.

10. Gardner and Gardner, *The Structure of Learning*, p. 296.

11. Ibid.

12. Ibid.

13. Fouts and Mills, *Next of Kin*, p. 30.

14. M. D. Bodamar and R. Allen, How Cross-Fostered Chimpanzees (Pan troglodytes) Initiate and Maintain Conversations. *Journal of Comparative Psychology* 116 (2002), 12–26. doi:10.1037/0735-7036.116.1.12.

15. Gardner and Gardner, *The Structure of Learning*, p. 298.

16. Fouts and Mills, *Next of Kin*, pp. 30–90.

17. Gardner and Gardner, *The Structure of Learning*, p. 294.

18. Ibid., p. 306.

19. Ibid., pp. 31–32.

20. J. Foer, The Truth about Chimps. *National Geographic*, February 2010.

21. Gardner and Gardner, *The Structure of Learning*, p. 291.

22. Fouts and Mills, *Next of Kin*, p. 30.

23. Bodamar and Allen, How Cross-Fostered Chimpanzees (Pan troglodytes) Initiate and Maintain Conversations.

24. Gardner and Gardner, *The Structure of Learning*, pp. 321–322.

25. R. Fouts and D. Fouts, Conversations with Chimpanzees: A Review of Recent Research, Research Methods and Enrichment Techniques. In

Proceedings of the Annual ChimpanZoo Conference, Individuality and Intelligence of Chimpanzees. Tucson, Arizona, 1995, pp. 51–53.
26. Fouts and Mills, *Next of Kin*, p. 291.

Chapter 6: No Language for a Lady: Gender and Swearing

1. K. Stapleton, Swearing. In M. Locher and S. Graham (eds), *Interpersonal Pragmatics* (Handbooks of Pragmatics 6). Mouton de Gruyter, 2010, pp. 289–306.
2. Peter Trudgill, *Sociolinguistics: An Introduction to Language and Society*, 4th ed. Penguin Books, 2000.
3. O. Jespersen, *Language; Its Nature, Development and Origin*. George Allen & Unwin, 1922.
4. E. Gordon, Sex, Speech, and Stereotypes: Why Women Use Prestige Speech Forms More Than Men. *Language in Society* 26 (1997), 47–63. doi:10.1017/S0047404500019400.
5. T. McEnery, *Swearing in English: Bad Language, Purity and Power from 1586 to the Present*. Routledge, 2006.
6. J. Collier, *A Short View of the Immorality and Profaneness of the English Stage; Together with the Sense of Antiquity upon this Argument*, 2nd ed. S. Keble, 1698.
7. V. de Klerk, How Taboo are Taboo Words for Girls? *Language in Society* 21 (1992), 277–289. doi:10.1017/S0047404500015293.
8. T. Jay, *Why We Curse: A Neuro-Psycho-Social Theory of Speech*. John Benjamins, 2000.
9. R. O'Neil, Sexual Profanity and Interpersonal Judgment. PhD dissertation, Louisiana State University, 2002.
10. T. Jay, The Utility and Ubiquity of Taboo Words. *Perspectives on Psychological Science* 4 (2009), 153–161. doi:10.1111/j.1745-6924.2009.01115.x.
11. McEnery, *Swearing in English*.
12. S. Holmes, Women Have Overtaken Men in Their Use of Profanities. *Mail Online* November 6 2016. www.dailymail.co.uk/news/article-3909524/Effing-Noras-swear-Normans-Women-overtaken-men-use-profanities.html.

13. L. A. Bailey and L. A. Timm, More on Women's—and Men's—Expletives. *Anthropological Linguistics* 18 (1976), 438–449. www.jstor.org/stable/30027592.

14. B. Risch, Women's Derogatory Terms for Men: That's Right, "Dirty" Words. *Language in Society* 16 (1987), 353–358. doi:10.1017/S0047404500012434.

15. de Klerk, How Taboo are Taboo Words for Girls?

16. S. E. Hughes, Expletives of Lower Working-Class Women. *Language in Society* 21 (1992), 291–303. doi:10.1017/S004740450001530X.

17. M. M. Oliver and J. Rubin, The Use of Expletives by Some American Women. *Anthropological Linguistics* 17 (1975), 191–197. www.jstor.org/stable/30027568.

18. M. Thelwall, *Fk Yea I Swear: Cursing and Gender in a Corpus of Myspace Pages.* Corpora, 2008.

19. E. Rassin and P. Muris, Why Do Women Swear? An Exploration of Reasons for and Perceived Efficacy of Swearing in Dutch Female Students. *Personality and Individual Differences* 38 (2005), 1669–1674. doi:10.1016/j.paid.2004.09.022.

20. Bailey and Timm, More on Women's – and Men's – Expletives.

21. C. Berger, *The Myth of Gender-Specific Swearing: A Semantic and Pragmatic Analysis.* Verlag für Wissenschaft und Forschung, Berlin, 2002.

22. Hughes, Expletives of Lower Working-Class Women.

23. E. Byrne and D. Corney, Sweet FA: Sentiment, Soccer and Swearing. In S. P. Papadopoulos, D. A. Cesar, A. Shamma, Kelliher, and R. Jain (eds.): *Proceedings of the SoMuS ICMR 2014 Workshop, Glasgow, Scotland,* 01-04-2014, published at http://ceur-ws.org.

24. A. Montagu, *The Anatomy of Swearing.* University of Pennsylvania Press, 1967.

Chapter 7: *Schieße, Merde, Cachau*: Swearing in Other Languages

1. J. Dewaele, Investigating the Psychological and Emotional Dimensions in Instructed Language Learning: Obstacles and Possibilities. *Modern Language Journal* 89 (2005), 367–380. doi:10.1111/j.1540-4781.2005.00311.x.

2. J. Dewaele, Blistering Barnacles! What Language Do Multilinguals Swear In?! *Estudios de Sociolingüística* 5 (2004), 83–105, at pp. 84–85.

3. E. M. Rintell, But How Did You FEEL About That? The Learner's Perception of Emotion in Speech. *Applied Linguistics* 5 (1984), 255–264. doi:10.1093/applin/5.3.255.

4. C. L. Harris, A. Ayçiçeği, and J. B. Gleason, Taboo Words and Reprimands Elicit Greater Autonomic Reactivity in a First Language Than in a Second Language. *Applied Psycholinguistics* 24 (2003), 561–579. doi:10.1017/S0142716403000286.

5. S. G. Kellman, *The Translingual Imagination*. University of Nebraska Press, 2000.

6. S. Ervin, Language and TAT Content in Bilinguals. *Journal of Abnormal Psychology* 68 (1964), 500–7.

7. S. Ervin-Tripp, An Analysis of the Interaction of Language, Topic, and Listener. *American Anthropologist* 66 (1964), 86–102.

8. J. Dewaele, The Emotional Force of Swearwords and Taboo Words in the Speech of Multilinguals. *Journal of Multilingual and Multicultural Development* 25 (2004), 204–222. doi:10.1080/01434630408666529.

9. J. Altarriba and D. M. Basnight-Brown, The Representation of Emotion vs. Emotion-Laden Words in English and Spanish in the Affective Simon Task. *International Journal of Bilingualism* 15 (2010), 310–328. doi:10.1177/1367006910379261.

10. A. Ayçiçeği and C. Harris, BRIEF REPORT Bilinguals' Recall and Recognition of Emotion Words. *Cognition and Emotion* 18 (2004), 977–987. doi:10.1080/02699930341000301.

11. Kellman, *The Translingual Imagination*.

12. Dewaele, The Emotional Force of Swearwords.

13. Ibid.

14. Dewaele, Blistering Barnacles!

15. Dewaele, The Emotional Force of Swearwords.

16. M. M. Chavez, The Orientation of Learner Language Use in Peer Work: Teacher Role, Learner Role and Individual Identity. *Language Teaching Research* 11 (2007), 161–188. doi:10.1177/1362168807074602.

17. Dewaele, Investigating the Psychological and Emotional Dimensions in Instructed Language Learning.

18. R.-E. Mercury, Swearing: A "Bad" Part of Language; A Good Part of Language Learning. *TESL Canada Journal* 13 (1995), 28–36. doi:10.18806/tesl.v13i1.659.

19. B. Rampton, Dichotomies, Difference, and Ritual in Second Language Learning and Teaching. *Applied Linguistics* 20 (1999), 316–340. doi:10.1093/applin/20.3.316.

20. W. McMorran, We Translated the Marquis de Sade's Most Obscene Work – Here's How. *Independent*, November 2, 2016. www.independent.co.uk/arts-entertainment/books/we-translated-the-marquis-de-sades-most-obscene-work-heres-how-a7393066.html.

21. A. M. Fernández Dobao, Linguistic and Cultural Aspects of the Translation of Swearing: The Spanish Version of *Pulp Fiction*. *Babel* 52 (2006), 222–242. doi:10.1075/babel.52.3.02fer.

22. J. Green, An Experiment in English. *Harper's Magazine*, September 1941, pp. 397–405.

23. Dewaele, Blistering Barnacles!

Bibliography

Ackermans, L., Duits, A., Linden, C. van der, Tijssen, M., Schruers, K., Temel, Y., Kleijer, M., Nederveen, P., Bruggeman, R., Tromp, S., Kranen-Mastenbroek, V. van, Kingma, H., Cath, D. and Visser-Vandewalle, V. Double-Blind Clinical Trial of Thalamic Stimulation in Patients with Tourette Syndrome. *Brain* 134 (2011), 832–844. doi:10.1093/brain/awq380.

Altarriba, J., and Basnight-Brown, D. M. The Representation of Emotion vs. Emotion-Laden Words in English and Spanish in the Affective Simon Task. *International Journal of Bilingualism* 15 (2010), 310–328. doi:10.1177/1367006910379261.

Aslaksen, P. M., Myrbakk, I. N., Høifødt, R. S., and Flaten, M. A. The Effect of Experimenter Gender on Autonomic and Subjective Responses to Pain Stimuli. *Pain* 129 (2007), 260–268. doi:10.1016/j.pain.2006.10.011.

Ayçiçeği, A., and Harris, C. BRIEF REPORT Bilinguals' Recall and Recognition of Emotion Words. *Cognition and Emotion* 18 (2004), 977–987. doi:10.1080/02699930341000301.

Bailey, L. A., and Timm, L. A. More on Women's—and Men's—Expletives. *Anthropological Linguistics* 18 (1976), 438–449. www.jstor.org/stable/30027592.

Baumeister, R. F., Vohs, K. D., and Tice, D. M. The Strength Model of Self-Control. *Current Directions in Psychological Science* 16 (2007), 351–355. doi:10.1111/j.1467-8721.2007.00534.x.

Bergen, B. K. *What the F: What Swearing Reveals about Our Language, Our Brains, and Ourselves.* Basic Books, 2016.

Berger, C. *The Myth of Gender-Specific Swearing: A Semantic and Pragmatic Analysis.* Verlag für Wissenschaft und Forschung, Berlin, 2002.

Bernstein, M. J., and Claypool, H. M. Social Exclusion and Pain Sensitivity: Why Exclusion Sometimes Hurts and Sometimes Numbs. *Personality and Social Psychology Bulletin* 38 (2012), 185–196. doi:10.1177/0146167211422449.

Black, K. J., Jankovic, J., Hershey, T., McNaught, K. S. P., Mink, J. W., and Walkup, J. Progress in Research on Tourette Syndrome. *Journal of Obsessive-Compulsive and Related Disorders* 3 (2014), 359–362. doi:10.1016/j.jocrd.2014.03.005.

Bloch, M. H., and Leckman, J. F. Clinical Course of Tourette Syndrome. *Journal of Psychosomatic Research* 67 (2009), 497–501. doi:10.1016/j.jpsychores.2009.09.002.

Bodamar, M. D., and Allen, R. How Cross-Fostered Chimpanzees (Pan troglodytes) Initiate and Maintain Conversations. *Journal of Comparative Psychology* 116 (2002), 12–26. doi:10.1037/0735-7036.116.1.12.

Byrne, E., and Corney, D. Sweet FA: Sentiment, Soccer and Swearing. In S. Papadopoulos, P. Cesar, D. A. Shamma, A. Kelliher, and R. Jain (eds.): *Proceedings of the SoMuS ICMR 2014 Workshop, Glasgow, Scotland, 01-04-2014,* published at http://ceur-ws.org.

Chavez, M. M. The Orientation of Learner Language Use in Peer Work: Teacher Role, Learner Role and Individual Identity. *Language Teaching Research* 11 (2007), 161–188. doi:10.1177/1362168807074602.

Clempson, A., Dobson S., and Stern, Judith S. P7 Dentists Treating Tourette Syndrome. *Journal of Neurology, Neurosurgery and Psychiatry* 83 (2012). doi:10.1136/jnnp-2012-303538.24.

Collier, J. *A Short View of the Immorality and Profaneness of the English Stage; Together with the Sense of Antiquity upon this Argument,* 2nd edn. S. Keble, 1698.

Conelea, C. A., Franklin, S. A., and Woods, D. W. Tic, Tourettes, and Related Disorders. In R. J. R. Levesque (ed.), *Encyclopedia of Adolescence.* Springer New York, 2011, pp. 2976–2983.

Daly, N., Holmes, J., Newton, J., and Stubbe, M. Expletives as Solidarity
  Signals in FTAs on the Factory Floor. *Journal of Pragmatics* 36 (2004),
  945–964. doi:10.1016/j.pragma.2003.12.004.

de Klerk, V. How Taboo are Taboo Words for Girls? *Language in Society* 21
  (1992), 277–289. doi:10.1017/S0047404500015293.

Deckman, T., DeWall, C. N., Way, B., Gilman, R. and Richman, S. Can
  Marijuana Reduce Social Pain? *Social Psychological and Personality
  Science* 13 (2013), 60–68. doi:10.1177/1948550613488949.

Dewaele, J. Blistering Barnacles! What Language Do Multilinguals
  Swear In?! *Estudios de Sociolingüística* 5 (2004), 83–105. doi:10.1558/
  sols.v5i1.83.

———. The Emotional Force of Swearwords and Taboo Words in
  the Speech of Multilinguals. *Journal of Multilingual and Multicultural
  Development* 25 (2004), 204–222. doi:10.1080/01434630408666529.

———. Investigating the Psychological and Emotional
  Dimensions in Instructed Language Learning: Obstacles
  and Possibilities. *Modern Language Journal* 89 (2005), 367–380.
  doi:10.1111/j.1540-4781.2005.00311.x.

DeWall, N. C., Macdonald, G., Webster, G. D., Masten, C. L.,
  Baumeister, R. F., Powell, C., Combs, D., Schurtz, D. R., Stillman,
  T. F., Tice, D. M., and Eisenberger, N. I. Acetaminophen Reduces
  Social Pain: Behavioral and Neural Evidence. *Psychological Science* 21
  (2010), 931–937. doi:10.1177/0956797610374741.

Donovan, J. M., and Anderson, H. E. *Anthropology and Law.* Berghahn
  Books, 2005.

Duits, A., Ackermans, L., Cath, D., and Visser-Vandewalle, V.
  Unfavourable Outcome of Deep Brain Stimulation in a Tourette
  Patient with Severe Comorbidity. *European Child and Adolescent
  Psychiatry* 21 (2012), 529–531.

Elliott, R. Executive Functions and Their Disorders: Imaging in Clinical
  Neuroscience. *British Medical Bulletin* 65 (2003), 49–59. doi:10.1093/
  bmb/65.1.49.

Ervin, S. Language and TAT Content in Bilinguals. *Journal of Abnormal
  Psychology* 68 (1964), 500–507.

Ervin-Tripp, S. An Analysis of the Interaction of Language, Topic, and
  Listener. *American Anthropologist* 66 (1964), 86–102.

Feldman, G., Lian, H., Kosinski M., and Stillwell, D. Frankly, We Do
  Give a Damn: The Relationship between Profanity and Honesty.

Social Psychological and Personality Science (published online January 2017). doi:10.1177/1948550616681055.

Fernández Dobao, A. M. Linguistic and Cultural Aspects of the Translation of Swearing: The Spanish Version of Pulp Fiction. *Babel* 52 (2006), 222–242. doi:10.1075/babel.52.3.02fer.

Foer, J. The Truth about Chimps. *National Geographic*, February 2010.

Fouts, R. and Fouts, D. Conversations with Chimpanzees: A Review of Recent Research, Research Methods, and Enrichment Techniques. In *Proceedings of the Annual ChimpanZoo Conference, Individuality and Intelligence of Chimpanzees*. Tucson, Arizona, 1995, pp. 51–53.

Fouts, R. and Mills, S. *Next of Kin: What My Conversations with Chimpanzees Have Taught Me about Intelligence, Compassion and Being Human.* Michael Joseph, 1997.

Fox, Kate. *Watching the English: The Hidden Rules of English Behaviour.* Hodder & Stoughton, 2004.

Gainotti, G. Unconscious Processing of Emotions and the Right Hemisphere. *Neuropsychologia* 50 (2012), 205–218. doi:10.1016/j .neuropsychologia.2011.12.005.

Gardner, R. A., and Gardner, B. *The Structure of Learning: From Sign Stimuli to Sign Language.* Lawrence Erlbaum, 1998.

Gordon, E. Sex, Speech, and Stereotypes: Why Women Use Prestige Speech Forms More Than Men. *Language in Society* 26 (1997), 47–63. doi:10.1017/S0047404500019400.

Green, J. An Experiment in English. *Harper's Magazine*, September 1941, pp. 397–405.

Harris, C. L., Ayçiçeği, A., and Gleason, J. B. Taboo Words and Reprimands Elicit Greater Autonomic Reactivity in a First Language Than in a Second Language. *Applied Psycholinguistics* 24 (2003), 561–579. doi:10.1017/S0142716403000286.

Haugh, M., and Bousfield, D. Mock Impoliteness, Jocular Mockery and Jocular Abuse in Australian and British English. *Journal of Pragmatics* 44 (2012), 1099–1114. doi:10.1016/j.pragma.2012.02.003.

Heath, R. L., and Blonder, L. X. Spontaneous Humor among Right Hemisphere Stroke Survivors. *Brain and Language* 93 (2005), 267–276. doi:10.1016/j.bandl.2004.10.006.

Hitchcock, E., and Cairns, V. Amygdalotomy. *Postgraduate Medical Journal* 49 (1973), 894–904. doi:10.1136/pgmj.49.578.894.

Holmes, S. Women Have Overtaken Men in Their Use of Profanities. *Mail Online* November 6 2016. www.dailymail.co.uk/news/article-3909524/Effing-Noras-swear-Normans-Women-overtaken-men-use-profanities.html.

Hughes, G. *Swearing: A Social History of Foul Language, Oaths and Profanity in English.* Blackwell, 1991.

Hughes, S. E. Expletives of Lower Working-Class Women. *Language in Society* 21 (1992), 291–303. doi:10.1017/S004740450001530X.

Indersmitten, T., and Gur, R. C. Emotion Processing in Chimeric Faces: Hemispheric Asymmetries in Expression and Recognition of Emotions. *Journal of Neuroscience* 23 (2003), 3820–3825.

Jay, K. L. and Jay, T. B. Taboo Word Fluency and Knowledge of Slurs and General Pejoratives: Deconstructing the Poverty-of-Vocabulary Myth. *Language Sciences* 52 (2015), 251–259. doi:10.1016/j.langsci.2014.12.003.

Jay, T. *Why We Curse: A Neuro-Psycho-Social Theory of Speech.* John Benjamins, 2000.

———. The Utility and Ubiquity of Taboo Words. *Perspectives on Psychological Science* 4 (2009), 153–161. doi:10.1111/j.1745-6924.2009.01115.x.

Jespersen, O. *Language; Its Nature, Development and Origin.* George Allen & Unwin, 1922.

Kandel, E., Schwartz, J., and Jessell, T. *Principles of Neural Science.* McGraw-Hill Medical, 2000.

Kellman, Steven G. *The Translingual Imagination.* University of Nebraska Press, 2000.

Kellogg, W., and Kellogg, L. *The Ape and the Child.* Hafner Publishing, 1933.

Kurlan, R. M. Treatment of Tourette Syndrome. *Neurotherapeutics* 11 (2013), 161–165. doi:10.1007/s13311-013-0215-4.

Lang, A. E., Consky, E., and Sandor, P. "Signing Tics"—Insights into the Pathophysiology of Symptoms in Tourette's Syndrome. *Annals of Neurology* 33 (1993), 212–215. doi:10.1002/ana.410330212.

Lepper, M. R., Greene, D., and Nisbett, R. E. Undermining Children's Intrinsic Interest with Extrinsic Reward: A Test of the "Overjustification" Hypothesis. *Journal of Personality and Social Psychology* 28 (1973), 129–137. doi:10.1037/h0035519.

Lombardo, L. Hurt Feelings and Four Letter Words: The Effects of Verbal Swearing on Social Pain. Honors Thesis, School of Psychology, The University of Queensland, 2012.

Lowe, N. K. The Nature of Labor Pain. *American Journal of Obstetrics and Gynecology* 186 (2002), S16–S24. doi:10.1016/S0002-9378(02)70179-8.

Maciunas, R. J., Maddux, B. N., Riley, D. E., Whitney, C. M., Schoenberg, M. R., Ogrocki, P. J., Albert, J. M., and Gould, D. J. Prospective Randomized Double-Blind Trial of Bilateral Thalamic Deep Brain Stimulation in Adults with Tourette Syndrome. *Journal of Neurosurgery* 107 (2007), 1004–1014. doi:10.3171/JNS-07/11/1004.

McEnery, T. *Swearing in English: Bad Language, Purity and Power from 1586 to the Present.* Routledge, 2006.

McMorran, W. We Translated the Marquis de Sade's Most Obscene Work—Here's How. *Independent*, 2 November 2016. www.independent.co.uk/arts-entertainment/books/we-translated-the-marquis-de-sades-most-obscene-work-heres-how-a7393066.html.

Mercury, R.-E. Swearing: A "Bad" Part of Language; A Good Part of Language Learning. *TESL Canada Journal* 13 (1995), 28–36. doi:10.18806/tesl.v13i1.659.

Michael, R. P. Treatment of a Case of Compulsive Swearing. *British Medical Journal* 1 (1957), 1506–1508.

Mohr, M. *Holy Sh*t: A Brief History of Swearing.* Oxford University Press, 2013.

Montagu, A. *The Anatomy of Swearing.* University of Pennsylvania Press, 1967.

Morris, D. *The Biology of Art: A Study of the Picture-Making Behaviour of the Great Apes and Its Relationship to Human Art.* Methuen, 1962.

Newman, M. L., Pennebaker, J. W., Berry, D. S., and Richards, J. M. Lying Words: Predicting Deception from Linguistic Styles. *Personality and Social Psychology Bulletin* 29 (2003), 665–675. doi:10.1177/0146167203029005010.

O'Connor, James V. *Cuss Control: The Complete Book on How to Curb Your Cursing.* Three Rivers Press, 2000.

O'Neil, R. Sexual Profanity and Interpersonal Judgment. PhD dissertation, Louisiana State University, 2002.

Ofcom. Attitudes to Potentially Offensive Language and Gestures on TV and Radio. www.ofcom.org.uk/research-and-data/tv-radio-and-on-demand/tv-research/offensive-language-2016.

Öhman, A., Flykt, A., Esteves, F., and Institut, K. Emotion Drives Attention: Detecting the Snake in the Grass. *Journal of Experimental Psychology: General* 130 (2001), 466–478. doi:10.1037/AXJ96-3445.130.3.466.

Okun, M. S., Foote, K. D., and Wu, S. S. A Trial of Scheduled Deep Brain Stimulation for Tourette Syndrome: Moving Away from Continuous Deep Brain Stimulation Paradigms. *Archives of Neurology* 70 (2013), 85–94. doi:10.1001/jamaneurol.2013.580.

Oliver, M. M., and Rubin, J. The Use of Expletives by Some American Women. *Anthropological Linguistics* 17 (1975), 191–197. www.jstor.org/stable/30027568.

Palminteri, S., Lebreton, M., Worbe, Y., Hartmann, A., Lehéricy, S., Vidailhet, M., Grabli, D., and Pessiglione, M. Dopamine-Dependent Reinforcement of Motor Skill Learning: Evidence from Gilles de la Tourette Syndrome. *Brain* 134 (2011), 2287–2301. doi:10.1093/brain/awr147.

Pavlenko, A. Bilingualism and Emotions. *Multilingua—Journal of Cross-Cultural and Interlanguage Communication* 21 (2002), 45–78. doi:10.1515/mult.2002.004.

Piacentini J., Woods D. W., Scahill L., Wilhelm, S., Peterson, A. L., Chang, S., Ginsburg, G. S., Deckersbach, T., Dziura, J., Levi-Pearl, S., and Walkup, J. T. Behavior Therapy for Children with Tourette Disorder: A Randomized Controlled Trial. *Journal of the American Medical Association* 303 (2010), 1929–1937. doi:10.1001/jama.2010.607.

Pinker, S. What the F***? *New Republic*, October 8, 2007. https://newrepublic.com/article/63921/what-the-f.

Plester, B., and Sayer, J. "Taking the Piss": Functions of Banter in the IT Industry. *Humor* 20 (2007), 157–187. doi:10.1515/HUMOR.2007.008.

Rampton, B. Dichotomies, Difference, and Ritual in Second Language Learning and Teaching. *Applied Linguistics* 20 (1999), 316–340. doi:10.1093/applin/20.3.316.

Rassin, E. and Muris, P. Why Do Women Swear? An Exploration of Reasons for and Perceived Efficacy of Swearing in Dutch Female Students. *Personality and Individual Differences* 38 (2005), 1669–1674. doi:10.1016/j.paid.2004.09.022.

Rassin, E., and van der Heijden, S. Appearing Credible? Swearing Helps! *Psychology, Crime and Law* 11 (2005), 177–182. doi:10.1080/1068316051605123319329952.

Ratiu, P., and Talos, I.-F. The Tale of Phineas Gage, Digitally Remastered. *New England Journal of Medicine* 351 (2004), e21. doi:10.1056/NEJMicmo31024.

Rayner, G. Sir Winston Churchill Quotes: The Famous Lines That He Never Said. *Telegraph*, 13 October 2014. www.telegraph.co.uk/news/politics/conservative/11155416/Sir-Winston-Churchill-the-famous-lines-that-he-never-said.html.

Rechtspraak.nl (database of Dutch court cases and rulings).

Rhudy, J. L., and Meagher, M. W. Negative Affect: Effects on an Evaluative Measure of Human Pain. *Pain* 104 (2003), 617–626. doi:10.1016/S0304-3959(03)00119-2.

Rintell, E. M. But How Did You FEEL About That? The Learner's Perception of Emotion in Speech. *Applied Linguistics* 5 (1984), 255–264. doi:10.1093/applin/5.3.255.

Risch, B. Women's Derogatory Terms for Men: That's Right, "Dirty" Words. *Language in Society* 16 (1987), 353–358. doi:10.1017/S0047404500012434.

Robbins, M. L., Focella, E. S., Kasle, S., López, A. M., Weihs, K. L., and Mehl, M. R. Naturalistically Observed Swearing, Emotional Support, and Depressive Symptoms in Women Coping with Illness. *Health Psychology* 30 (2011), 789–792. doi:10.1037/a0023431.

Romanes, G. J. *Mental Evolution in Man*. D. Appleton and Co., 1889.

Saisto, T., Kaaja, R., Ylikorkala, O., and Halmesmäki, E. Reduced Pain Tolerance during and after Pregnancy in Women Suffering from Fear of Labor. *Pain* 93 (2001), 123–127. doi:10.1016/S0304-3959(01)00302-5.

Scherer, C., and Sagarin, B. Indecent Influence: The Positive Effects of Obscenity on Persuasion. *Social Influence* 1 (2006), 138–146. doi:10.1080/15534510600747597.

Seymour-Smith, S. Illness as an Occasion for Storytelling: Social Influences in Narrating the Masculine Self to an Unseen Audience. In C. Horrocks, K. Milnes and G. Roberts (eds), *Narrative, Memory and Life Transitions*. Huddersfield University Press, 2002.

———. "Blokes Don't Like That Sort of Thing": Men's Negotiation of a "Troubled" Self-help Group Identity. *Journal of Health Psychology* 13 (2008), 785–797. doi:10.1177/1359105308093862.

Shammi, P., and Stuss, D. T. Humour Appreciation: A Role of the Right Frontal Lobe. *Brain* 122 (1999), 657–666. doi:10.1093/brain/122.4.657.

Smith, H., Fox, J. R. E., Hedderly, T., Murphy, T., and Trayner, P. Investigating Young People's Experiences of Successful or Helpful Psychological Interventions for Tic Disorders: An Interpretative Phenomenological Analysis Study. *Journal of Health Psychology* 21 (2016), 1787–1798. doi:10.1177/1359105314566647.

Speedie, L. J., Wertman, E., Ta'ir, J., and Heilman, K. M. Disruption of Automatic Speech following a Right Basal Ganglia Lesion. *Neurology* 43 (1993), 1768–1768. doi:10.1212/WNL.43.9.1768.

Stapleton, K. Swearing. In M. Locher and S. Graham (eds), *Interpersonal Pragmatics* (Handbooks of Pragmatics 6). Mouton de Gruyter, 2010, pp. 289–306.

Stephens, R., and Allsop, C. Effect of Manipulated State Aggression on Pain Tolerance. *Psychological Reports* 111 (2012), 311–321. doi:10.2466/16.02.20.

Stephens, R., Atkins, J., and Kingston, A. Swearing as a Response to Pain. *Neuroreport* 20 (2009), 1056–1060. doi:10.1097/WNR.0b013e32832e64b1.

Stephens, R., and Umland, C. Swearing as a Response to Pain—Effect of Daily Swearing Frequency. *Journal of Pain* 12 (2011), 1274–1281. doi:10.1016/j.jpain.2011.09.004.

Stroop, J. R. Studies of Interference in Serial Verbal Reactions. *Journal of Experimental Psychology* 18 (1935), 643–662. doi:10.1037/h0054651.

Sylwester, K., and Purver, M. Twitter Language Use Reflects Psychological Differences between Democrats and Republicans. *PLoS ONE* 10, 2015, e0137422. doi:10.1371/journal.pone.0137422.

Terrace, H. S. *Nim: A Chimpanzee Who Learned Sign Language.* Columbia University Press, 1986.

Thelwall, M. *Fk Yea I Swear: Cursing and Gender in a Corpus of Myspace Pages.* Corpora, 2008.

Trudgill, Peter. *Sociolinguistics: An Introduction to Language and Society*, 4th ed. Penguin Books, 2000.

Van Lancker, D., and Cummings, J. Expletives: Neurolinguistic and Neurobehavioral Perspectives on Swearing. *Brain Research Reviews* 31 (1999), 83–104. doi:10.1016/S0165-0173(99)00060-0.

Van Lancker, D., and Klein, K. Preserved Recognition of Familiar Personal Names in Global Aphasia. *Brain and Language* 39 (1990), 511–529. doi:10.1016/0093-934X(90)90159-E.

Wadman, R., Tischler, V., and Jackson, G. M. "Everybody just thinks I'm weird": A Qualitative Exploration of the Psychosocial Experiences of Adolescents with Tourette Syndrome. *Child Care Health and Development* 39 (2013), 880–886. doi:10.1111/cch.12033.

Welhoffer, E. Strafe Für Beleidigungen: Wie Teuer Ist Der "Stinkefinger"? *Express.de*, 15 March 2016. www.express.de/news/politik-und-wirtschaft/recht/beleidigung-beschimpft-strafe-schimpfwort-teuer-anzeige-strafbar-1261268-seite2.

Wilhelm, S., Peterson, A. L., Piacentini, J., Woods, D. W., Deckersbach, T., Sukhodolsky, D. G., Chang, S., Liu, H., Dziura, J., Walkup, J. T., and Scahill, L. Randomized Trial of Behavior Therapy for Adults with Tourette Syndrome. *Archives of General Psychiatry* 69 (2012), 795–803. doi:10.1001/archgenpsychiatry.2011.1528.

Williams, A. E. and Rhudy, J. L. Emotional Modulation of Autonomic Responses to Painful Trigeminal Stimulation. *International Journal of Psychophysiology* 71 (2009), 242–247. doi:10.1016/j.ijpsycho.2008.10.004.

Yerkes, R. M. *Almost Human*. Jonathan Cape, 1925.

Index

INDEX